DEVELOPMENTAL REGULATION

Aspects of Cell Differentiation

CELL BIOLOGY: A Series of Monographs

EDITORS

D. E. BUETOW

*Department of Physiology
and Biophysics
University of Illinois
Urbana, Illinois*

I. L. CAMERON

*Department of Anatomy
University of Texas
Medical School at San Antonio
San Antonio, Texas*

G. M. PADILLA

*Department of Physiology and Pharmacology
Duke University Medical Center
Durham, North Carolina*

G. M. Padilla, G. L. Whitson, and I. L. Cameron (editors). THE CELL CYCLE: Gene-Enzyme Interactions, 1969

A. M. Zimmerman (editor). HIGH PRESSURE EFFECTS ON CELLULAR PROCESSES, 1970

I. L. Cameron and J. D. Thrasher (editors). CELLULAR AND MOLECULAR RENEWAL IN THE MAMMALIAN BODY, 1971

I. L. Cameron, G. M. Padilla, and A. M. Zimmerman (editors). DEVELOPMENTAL ASPECTS OF THE CELL CYCLE, 1971

P. F. Smith. THE BIOLOGY OF MYCOPLASMAS, 1971

Gary L. Whitson (editor). CONCEPTS IN RADIATION CELL BIOLOGY, 1972

Donald L. Hill. THE BIOCHEMISTRY AND PHYSIOLOGY OF *TETRA-HYMENA*, 1972

Kwang W. Jeon (editor). THE BIOLOGY OF AMOEBA, 1973

Dean F. Martin and George M. Padilla (editors). MARINE PHARMACOGNOSY: Action of Marine Biotoxins at the Cellular Level, 1973

Joseph A. Erwin (editor). LIPIDS AND BIOMEMBRANES OF EUKARYOTIC MICROORGANISMS, 1973

A. M. Zimmerman, G. M. Padilla, and I. L. Cameron (editors). DRUGS AND THE CELL CYCLE, 1973

Stuart Coward (editor). DEVELOPMENTAL REGULATION: Aspects of Cell Differentiation, 1973

In preparation

Govindjee (editor). BIOENERGETICS OF PHOTOSYNTHESIS

DEVELOPMENTAL REGULATION

Aspects of Cell Differentiation

Edited by STUART J. COWARD

Department of Zoology
University of Georgia
Athens, Georgia

1973

ACADEMIC PRESS   New York and London

A Subsidiary of Harcourt Brace Jovanovich, Publishers

ACADEMIC PRESS, INC.
111 Fifth Avenue, New York, New York 10003

United Kingdom Edition published by
ACADEMIC PRESS, INC. (LONDON) LTD.
24/28 Oval Road, London NW1

LIBRARY OF CONGRESS CATALOG CARD NUMBER: 71-187254

PRINTED IN THE UNITED STATES OF AMERICA

Contents

8.   Some Comparative Aspects of Cardiac and Skeletal Myogenesis

Francis J. Manasek

9.   Chondrogenesis

Robert Searls

List of Contributors

Numbers in parentheses indicate the pages on which the authors' contributions begin.

W. Sue Badman, Society for Developmental Biology, Kalamazoo, Michigan (85)

Stuart Brody, Department of Biology, University of California at San Diego, La Jolla, California (107)

Leon S. Dure, III, Department of Biochemistry, University of Georgia, Athens, Georgia (23)

James H. Gregg, Department of Zoology, University of Florida, Gainesville, Florida (85)

Tom Humphreys, Kewalo Marine Laboratory, Pacific Biomedical Research Center, University of Hawaii, Honolulu, Hawaii (1)

Joe L. Key, Department of Botany, University of Georgia, Athens, Georgia (49)

Gary Kochert, Department of Botany, University of Georgia, Athens, Georgia (155)

Thomas L. Lentz, Department of Anatomy, Yale University School of Medicine, New Haven, Connecticut (169)

Francis J. Manasek, Departments of Anatomy and Pediatrics, Harvard Medical School, Boston, Massachusetts, and Departments of Cardiology and Pathology, The Children's Hospital Medical Center, Boston, Massachusetts (193)

Robert Searls, Department of Biology, Temple University, Philadelphia, Pennsylvania (219)

Larry N. Vanderhoef, Department of Botany, University of Illinois, Urbana, Illinois (49)

Preface

In a field of such wide scope as developmental biology, it is impossible to include or to do justice to each and every important facet in any one book. I have attempted to draw together in this work some studies which are directed toward the orderly changes in cell phenotypes which we understand to be developmental in nature. The exciting, fundamental advances in genetics, cell biology, and molecular biology occurring over the last decade not only have reshaped the investigator's thinking and reinforced his armamentarium, but have brought developmental studies to a critical threshold point toward the elucidation of the basic mechanisms of differentiation. These changes, I believe, are reflected in the nine chapters of this volume. The linear ordering of the chapters obviously should not be regarded as describing the principal logical interactions. This would require arrangement within a spheroid; it is for the reader to make that juxtaposition.

STUART J. COWARD

1

RNA and Protein Synthesis during Early Animal Embryogenesis

TOM HUMPHREYS

I. Introduction

Embryogenesis is an orderly progression of the organism from a relatively simple fertilized egg to a much more complex functional individual. The rather similar cells resulting from cleavage must differentiate into a host of different cell types organized into diverse tissues and specialized for a variety of functions. Modern analysis of cellular biochemistry and molecular biology has established a scheme of molecular function and interdependency which leads to the general assertion that the central event in this morphological diversification is the orderly synthesis of appropriate enzymatic, structural, and regulatory proteins. It is the proteins, with their extensive molecular diversity, which perform most cellular functions. RNA molecules copied from the DNA encoding the protein's amino acid sequence direct the assembly and synthesis of the protein molecules. The activities of the DNA and RNA which specify the structure of the proteins appear to be regulated by protein molecules. The other molecules of cells, such as polysaccharides and lipids, are themselves synthesized and usually organized by protein mole-

1

cules. Thus, the sequence of synthesis of the many specific RNA and protein molecules required for development and the mechanisms regulating this sequence are at the crux of understanding development. Current studies are actively pursuing this basic question. The sequence of synthesis of some major classes of molecules has been worked out, and in a few cases the sequence for defined molecules has been described. Aspirations to penetrate beyond this gross description to reveal the molecular regulatory mechanisms which may control synthesis have not yet been fulfilled.

In this review I will attempt to provide a concise overview of the studies on molecular synthesis during early development. Since an extensive treatise has recently been published covering this topic (Davidson, 1969), I will not attempt to deal with the complete literature, but will emphasize areas in which I believe new insights have been gained and fresh viewpoints need to be expressed. I will also try to point beyond today's results to important questions which are within the reach of current technical capabilities. I will emphasize sea urchin embryos because they are probably the most studied and because I work on them and know them best.

II. Synthesis of RNA

A. RNA Synthesis before Fertilization

RNA synthesis in the mature oocyte, or egg, before fertilization has not been elucidated. It has been said that RNA synthesis in the unfertilized egg is at a reduced, basal level (for example, Davidson, 1969). The actual results available are fragmentary and divergent from different organisms, probably because the storage of fully grown oocytes is achieved differently in various species. This belief is most strongly supported by the apparent constant amount of ribosomal RNA with no turnover of rRNA (Davidson et al., 1964) during storage of oocytes in amphibians. However, rRNA synthesis does occur in some oocytes including amphibian (Gould, 1969a; Smith and Ecker, 1970), but its rate has not been measured.

In sea urchins, where the egg is stored after the meiotic divisions, the egg is impermeable to radioactive RNA precursors (Piatigorsky and Whiteley, 1965). I have attempted to set an upper limit for RNA synthesis in sea urchin eggs by measuring the specific activity of the precursor pool during labeling experiments as we have during embryogenesis (Emerson and Humphreys, 1971; Brandhorst and Humphreys, 1971).

Very high concentrations of radioactive precursor (50 μCi [^{3}H]adenosine/ml) were used in these experiments without achieving significant radioactivity in RNA. Sometimes, small amounts of radioactive ribosomal RNA were obtained but this may have been due to immature oocytes even though all preparations used contained more than 99.5% normal eggs. When the specific activities of the ATP pools were measured during these incubations, they were so low that RNA synthesis at the rate observed in eggs just fertilized would have gone undetected. The rate and kinds of RNA synthesized in unfertilized eggs could be identical to those of the eggs just fertilized. Since, as I will show below, RNA synthesis on a per nucleus basis is very active immediately after fertilization, failure to label RNA is no support for the idea that synthesis is reduced or at a basal level.

In the mature amphibian oocyte, heterogenous RNA synthesis appears to be rather rapid. Brown and Littna (1966) showed that a very large amount, considering the egg has one nucleus, of heterogenous RNA was radioactive in *Xenopus* eggs being shed by females injected with ^{32}P. The amount of RNA was constant through cleavage, and they believed that it was synthesized during ovulation and persisted. Recently Smith and Ecker (1970) have shown that RNA can be labeled at all stages of egg maturation. This finding suggests that the radioactive RNA observed by Brown and Littna is a steady state level of unstable RNA which is turning over and continually resynthesized and can become radioactive whenever precursor is added. Since the amount of this RNA, 10^{-9} gm per egg, is above two orders of magnitude more than is accumulated by a single nucleus in most animal cells (for example, see Brandhorst and Humphreys, 1971), I suggest that it might be mitochondrial RNA synthesized on the very large excess of mitochondrial DNA in the amphibian egg.

The activity of the ribosomal RNA genes is thought to be reduced in mature oocytes and eggs (Davidson, 1968) but again quantitative data is lacking. Although an absolute rate has not been measured, Smith and Ecker (1970) show labeled nucleoli in mature oocytes suggesting some rRNA synthesis. Calculations I have done suggest that the large amount of heterogenous RNA synthesized in mature oocytes could mask the activity of the large number of amplified ribosomal genes present at this stage even if they were transcribed at the rate observed in gastrulae (Brown and Littna, 1966). The full implications of not shutting down the ribosomal genes at the end of oogenesis are not clear to me, but in many amphibians such as *Rana pipiens*, it would make little difference since synthesis would stop as the temperature dropped in the fall after the eggs are fully grown. The same argument does not hold for *Xenopus*, but this problem has not been examined in enough detail to answer the question either way.

The best documented case of rRNA synthesis in full-grown oocytes is in *Urechis capo* (Gould, 1969a). All stages of these eggs are permeable and ribosomal RNA synthesis is easily detected. The data do not permit an estimate of the actual rates of synthesis, and the synthesis observed may be a residual, repressed level. It also might as easily be equal to synthesis in the young, growing oocytes. The evidence suggests that the ribosomal RNA synthesized in the mature oocytes of *Urechis* is not fully processed and does not enter ribosomes. It, therefore, might not contribute to the store of egg ribosomes. It would seem most logical that RNA synthesis, similar to some other metabolism including protein synthesis, is reduced in the storage oocyte or egg and is reactivated at ovulation, fertilization, or later. However, sound evidence to justify this conclusion is lacking and the details of the regulation of RNA synthesis at the end of oogenesis and at fertilization are yet to be worked out.

B. RNA SYNTHESIS DURING DEVELOPMENT

Many early embryos exhibit unique patterns of RNA synthesis which are not observed in other kinds of cells. With development, the patterns pass through a series of changes and finally revert to the pattern typical of most animal cells with a rapid synthesis of large nuclear RNA and a predominate accumulation of ribosomal and transfer RNA. These changes have been described best in sea urchin embryos (Emerson and Humphreys, 1970). Right after fertilization the RNA labeled during an incubation with radioactive RNA precursors is small heterogenous RNA which sediments from 5 to 25 S. Toward the end of exponential cell division the predominate RNA becomes larger and ranges up to 50 S or more. At about gastrulation, when the rapid cell divisions of early development are complete, ribosomal and transfer RNA become the predominate RNA's accumulating. Although the data are fragmentary, this pattern appears to occur in amphibians, teleosts, *Urechis*, insects, ascidians, and snails (Brown and Littna, 1964; Belitsina *et al.*, 1964; Gould, 1969b; Harris and Forrest, 1967; Markert and Cowden, 1965; Collier, 1966).

These changing patterns have been recognized for a decade but have only recently been understood. At present we may determine four separate stages of RNA synthesis. The initial period of small heterogenous RNA consists of two separate phases with different kinds of RNA predominating. In the first phase most of the RNA is synthesized from the mitochondrial DNA, which, because of the large cytoplasm to nucleus ratio, is the most abundant DNA in many eggs (Dawid, 1966; Piko et al., 1967). As nuclei increase in number during cleavage, their RNA synthesis predominates. However, during cleavage they appear to synthesize mostly small molecular weight heterogenous RNA (Emerson and Humphreys, 1970). Presumably this RNA codes for rapid nuclear protein synthesis which occurs during cleavage (Nemer and Lindsay, 1969; Kedes and Gross, 1969). As cleavage slows, the more typical large molecular weight RNA characteristic of nuclear RNA begins to predominate (Emerson and Humphreys, 1970). Accumulation of detectable, newly synthesized ribosomal RNA begins at about gastrulation when the synthesis of heterogenous RNA declines. In the following sections, I will summarize the data which establish the existence of these four periods of RNA synthesis.

Mouse and *Ascaris* embryos show little or no indication of these periods when heterogenous RNA predominates (Ellen and Gwatkin, 1968; Woodland and Graham, 1969; Piko, 1970). In these embryos ribosomal RNA predominates from very early stages. The reasons for the differences among early embryos is unknown, but the embryos which show the changes in patterns of RNA synthesis have rapid cleavage stages while those which do not show the changes cleave rather slowly.

1. *Mitochondrial RNA Synthesis*

Eggs contain tens to thousands of times more cytoplasm than the average diploid somatic cell. This cytoplasm contains mitochondria and thus much more mitochondrial DNA than is usual for a single cell. In sea urchin eggs at least 80% of the DNA is mitochondrial (Piko *et al.*, 1967). The amphibian egg DNA likewise contains a very large percentage of mitochondrial DNA (Dawid, 1966). In sea urchins this DNA appears to serve as a template for much of the RNA synthesized right after fertilization. The first data suggesting this came from experiments with artificially activated, enucleate half eggs. RNA synthesized in the enucleate halves was identical in quantity and kind to that synthesized in nucleate halves and whole eggs (Chamberlain, 1967, 1970). The nature of this RNA synthesized in the egg cytoplasm was established by showing that it hybridized well with purified sea urchin mitochondrial DNA (Craig, 1970) but not with DNA from sea urchin nuclei. The RNA synthesized during this time is heterogenous with a sharp peak at 15 to 17 S. It may be mitochondrial messenger RNA (Craig, 1970), but its nature and its role in mitochondrial physiology is essentially unexplored despite the fact that these early zygotes provide an ideal system for analyzing the details of mitochondrial RNA synthesis.

2. *Messenger RNA Synthesis during Cleavage*

Within a few cell divisions after fertilization, the size distribution of the newly synthesized RNA becomes slightly modified. The RNA still sediments in about the same size range but the distribution is broader and two peaks appear at about 21 and 10 S (Kedes and Gross, 1969). Much of the RNA synthesized at this time may be involved in synthesis of the nuclear proteins needed during the rapid cell divisions of cleavage (Nemer and Lindsay, 1969). The RNA in the 10 S peak has been characterized as at least several RNA species (Kedes and Gross, 1969) synthesized on reiterated nuclear genes (Kedes and Birnstiel, 1971). Further studies are required to establish the nature of most of this heterogenous RNA. Hybridization studies indicate that it still hybridizes well with mitochondrial DNA and a significant amount of it may still be mitochondrial (Hartman and Comb, 1969). More than likely, most of it is nuclear since mitochondrial RNA hybridizes much more rapidly with the small, mitochondrial genome than nuclear RNA does with the complex nuclear DNA.

Measurements I have made on rates of RNA synthesis during cleavage stages suggest that there is a constant rate of RNA synthesis until about the 16-cell stage and then the rate increases rapidly with increasing nuclei during cleavage. Figure 1 shows an accumulation curve for radioactive RNA when embryos are incubated with radioactive RNA precursor immediately after fertilization. An accumulation curve of this kind reflects both synthesis and decay (Emerson and Humphreys, 1971a; Brandhorst and Humphreys, 1971). If RNA is decaying rapidly in a steady state situation, new synthesis of radioactive RNA is soon balanced by the decay of radioactive RNA. Such an equilibrium is approached, as shown in the curve in Fig. 1, between 4 to 6 hr. After 6 hr the amount of radioactive RNA again increases rapidly indicating a nonequilibrium state. This rapid increase appears to continue as long as the cells are dividing. I interpret the first period of equilibrium as a time when the constant mitochondrial RNA synthesis predominates and the amount of RNA synthesized in nuclei is insignificant. This RNA is synthesized at about 2×10^{-13} gm/min/embryo with one nucleus before first cleavage. This rate may be compared with a rate of 3×10^{-15} gm/min/nucleus at pluteus stage (Brandhorst and Humphreys, 1971). The RNA turns over rapidly with a half-life on the order of 15 to 30 min (see Brandhorst and Humphreys, 1971, for full details for deriving this half-life from the data in Fig. 1). The rapid increase in RNA after the 16-cell stage reflects an increased rate of RNA synthesis as shown by the rapid initial accumulation of RNA when label is added

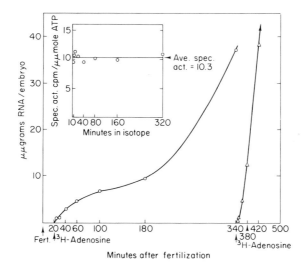

Fɪɢ. 1. Accumulation of radioactive RNA during an incubation of just-fertilized embryos (*Lytechinus pictus*) with [³H]adenosine. Ten μCi [³H]adenosine was added to a 5% (v/v) suspension 20 min after fertilization. Samples were taken at the time points indicated (circles), specific activity of the ATP and total radioactivity in RNA per embryo determined, and the amount of radioactive RNA calculated (Emerson and Humphreys, 1971; Brandhorst and Humphreys, 1971). [³H]Adenosine was added to a second batch of embryos at 340 min after fertilization and they were processed similarly (triangles).

at later times (Fig. 1). The increased rate of synthesis appears to parallel cleavage and probably is the simple result of exponential increase in number of nuclei during cleavage. Thus it would appear that mitochondrial RNA synthesis is rapidly diluted after the 16-cell stage when the nuclei double several times in the next few hours. Thus, I believe that probably the RNA synthesized during late cleavage is mostly nuclear although it is smaller than the typical nuclear RNA. However, this interpretation must be substantiated by more direct experiments characterizing both the nuclear and mitochondrial RNA species.

There are many questions that remain concerning this small RNA synthesized during late cleavage. Some of it moves to the polysomes and serves as messenger RNA (Kedes and Gross, 1969). How much of it does this and how much may decay in the nucleus is not known. Our techniques (Brandhorst and Humphreys, 1971, 1972) which measured the synthesis and decay rates of nuclear and cytoplasmic RNA at later stages (see below) cannot be applied during cleavage when rates of RNA synthesis are changing rapidly and a steady state condition does not exist. Careful examination of RNA metabolism at this stage for comparison with stages showing more typically sized nuclear RNA may provide some useful clues concerning the relationship of nuclear and messenger RNA.

3. *Heterogenous RNA Synthesis in Blastulae*

When cleavage is complete, the heterogenous RNA synthesized becomes much larger, ranging up to 50 S or more (Emerson and Humphreys, 1970). This RNA represents the heterogenous RNA typical of most eukaryotic cells, consisting of rapidly decaying nuclear RNA and the more stable messenger RNA (Brandhorst and Humphreys, 1972). In sea urchins at 18°C the nuclear RNA decays with half-life of 7 min and represents 85% of total synthesis and 35% of the steady state amount of heterogenous RNA. The cytoplasmic and polysomal messenger RNA decays with a half-life of 70 min and represents 15% of synthesis and 65% of the steady state level of heterogenous RNA. During this time ribosomal RNA represents much less than 1% of synthesis and only 10% of total accumulated RNA after a 10-hr labeling period (Emerson and Humphreys, 1970). RNA which sediments at 4 S to 5 S also accumulates in the cytoplasm during this time and may have been mistaken for stable species. Although part of this may be transfer RNA and 5 S ribosomal RNA, it labels as an unstable RNA species (from data in Brandhorst, 1971). The nature of this unstable RNA should be examined further (see Aronson, 1972).

4. *RNA Synthesis Postblastula*

As gastrulation begins, ribosomal RNA synthesis starts to predominate (Nemer, 1963; Brown and Littna, 1964; Comb *et al.*, 1965; Nemer and Infante, 1967; Giudice and Mutolo, 1967). This change in the pattern of RNA synthesis is due entirely to a reduction in the amount of heterogenous RNA which accumulates; ribosomal RNA continues to be synthesized at a constant rate (Kijima and Wilt, 1969; Emerson and Humphreys, 1970, 1971a). The reduction in accumulation of heterogenous RNA is caused by a coordinate reduction in the rate of synthesis of nuclear and messenger RNA without any change in the stability of these classes of RNA (Brandhorst and Humphreys, 1971, 1972).

5. *Regulation of Ribosomal RNA Synthesis*

There is little net accumulation of ribosomal RNA during early development of most species of embryos. In sea urchins there is about a 20% increase by the 72-hr pluteus stage (Fig. 2). However, as embryos begin to grow in later stages, they must accumulate more ribosomal RNA. Because ribosomal RNA synthesis does not predominate at early

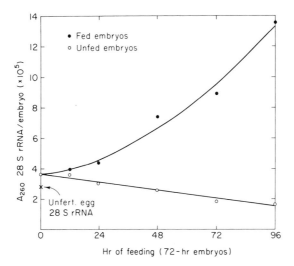

FIG. 2. 28 S ribosomal RNA in eggs and embryos either starved or fed at pluteus stage. *Lytechinus pictus* eggs were fertilized and allowed to develop in a gently shaken flask at a concentration of 100 eggs/ml sea water. When a functional gut had differentiated (72 to 90 hr), they were fed 2000 *Rhodomonas* (Hinegardner, 1969) per embryo. The seawater was changed every 12 hr and an equal number of algae were added again. Samples were taken at points indicated (filled and open circles), the embryos washed, and their RNA extracted with about 95% yield. The RNA was centrifuged on a sucrose gradient and the A_{260} in 28 S RNA determined (Humphreys, 1971).

stages but begins to do so at gastrulation, many workers became convinced that this change in RNA synthesis was correlated with the later changes in accumulation of ribosomal RNA. As noted above, our recent data from sea urchins shows a declining rate of heterogenous RNA synthesis which allows a constant rate of ribosomal RNA synthesis to become evident (Emerson and Humphreys, 1970, 1971a; Brandhorst and Humphreys, 1971, 1972). Measurements of RNA in amphibian embryos (Brown and Littna, 1966), when calculated on a per nucleus basis, showed that heterogenous RNA declined rapidly until about gastrulation and that the rate of ribosomal RNA synthesis was constant at all times it was measured. Biochemical data from other embryos are at best ambiguous; thus, the most complete studies on RNA synthesis make it seem unlikely that ribosomal RNA synthesis is activated at gastrulation in embryos.

The atypical nucleoli appearing in cleaving embryos and thought to be associated with repressed ribosomal RNA synthesis must be reinterpreted (Emerson and Humphreys, 1971a). We performed experiments which suggest that the multiple, small nucleoluslike structures in cleaving sea urchin nuclei can be correlated with rapid cell division. When cell division was stopped at the 16-cell stage these structures became much larger and more prominent. Although this does not settle the very interesting issue of the nature of these numerous prenucleoluslike bodies or the reason for their predominance over typical nucleoli in early stages, the results suggest that there is no compelling reason to correlate the prenucleolus-like bodies with reduction of ribosomal RNA synthesis.

Even though we became more and more convinced that there was no change in the rate of ribosomal RNA synthesis during early development, there were some disturbing issues which required examination. The rate of ribosomal RNA synthesis in cleavage or pluteus stage embryos was only 300 molecules/cell/hr (Emerson and Humphreys, 1970, 1971). Since a pluteus cell contains about 2×10^5 ribosomes, it would take about 25 to 30 days to double ribosomal RNA at this rate of synthesis. Obviously a growing cell must produce ribosomal RNA at a faster rate. However, it was not clear when sea urchin embryos began to grow. When we measured total ribosomal RNA in pluteus we found that it actually decreased as the plutei grew older (Fig. 2).

The idea which proved to be the key to this problem came from Hinegardner's (1969) experiment with feeding sea urchin larvae. His results indicated that they began to grow when they were fed. Dr. Hinegardner generously provided a culture of his *Rhodomonas* which were especially suitable for sea urchin larvae and I fed some embryos. Within 24 hr the embryos had significantly increased their ribosomal RNA (Fig. 2). and by 4 days ribosomal RNA had increased about four times. Labeling experiments confirmed that the rate of accumulation of new ribosomal RNA had increased. Within 5 hr after feeding the rate of synthesis of 28 and 18 S ribosomal RNA had increased fourfold (Fig. 3) and was even higher by 24 hr. Analysis of DNA and protein synthesis show that these are also stimulated by feeding. These results show that increase of ribosomal RNA synthesis is intimately linked with the initiation of growth in sea urchins and is regulated by feeding.

Further examination of this increased accumulation of ribosomal RNA should be very interesting. I presume it is achieved by increased synthesis. However, ribosomal RNA is synthesized as a precursor which is processed to yield the 28 and 18 S ribosomal RNA (Sconzo *et al.*, 1972). It is possible that accumulation of ribosomal RNA is regulated at the processing step since I have done no quantitative experiments on synthesis of precursor. These experiments will not be easy since the precursor in sea urchins is processed very rapidly and its pool is very small.

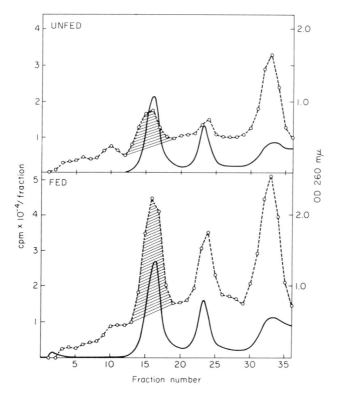

Fig. 3. The effect of feeding on the incorporation of radioactive adenosine into ribosomal RNA. *Lytechinus pictus* embryos (82-hr) were fed 2000 *Rhodomonas* per embryo and allowed to feed 5 hr. A sample of unfed and fed embryos were incubated 180 min with 5 μCi [^{3}H]adenosine (28 Ci/mmole)/ml and their RNA extracted. The RNA was layered onto sucrose–SDS gradients and centrifuged 14 hr at 24,000 rpm in a Beckman SW27 rotor at 30°C. The gradients were analyzed for A$_{260}$ (solid line) and trichloroacetic acid precipitable radioactivity (dashed line) (Emerson and Humphreys, 1970). Samples of embryos were taken at 40 and 180 min of labeling to determine the specific activity of the ATP pool (Emerson and Humphreys, 1971) which was about 3.7 $\times$ 10^{14} counts/min/mole ATP in all samples. The amount of radioactive 28 S ribosomal RNA (stripped area) was calculated to be 2.8 $\times$ 10^{-15} gm/unfed embryo and 14.3 $\times$ 10^{-15} gm/fed embryo.

Although we conclude that expression of the ribosomal genes is regulated by feeding, such a general conclusion is not very satisfying. This control must involve a series of specific reactions involving specific molecules. We would like to know what nutrient or nutrients are the most important. Might it be a single, small molecule such as an amino acid or nucleotide which we can add at any time to increase ribosomal RNA synthesis. The chain of control reactions must involve some intracellular regulator molecules and finally extend to the molecules which react with the genes themselves. Analysis of the molecular reactions at the level of the genes should be an especially promising area of study since both the genes (Patterson and Stafford, 1971) and the polymerases (Roeder and Rutter, 1969) for ribosomal RNA have been isolated from sea urchins.

The possibility that ribosomal RNA synthesis in amphibian and other embryos is regulated by feeding must be considered. Rapid accumulation of ribosomal RNA in *Rana pipiens* does not begin until about the feeding stages (Brown and Caston, 1962). Significant increase in total ribosomal RNA in *Xenopus* occurs only at late stages, when it is possible that feeding is occurring (Brown and Littna, 1964). Only further study can validate the generality of this possibility.

C. SUMMARY OF RNA SYNTHESIS

The present descriptions of RNA metabolism during development must be considered at best a meager beginning. The basic questions about the sequence of the expression of genes during development remains virtually untouched. Some experiments attempting to analyze this problem have been performed using DNA–RNA hybridization (Denis, 1966; Whiteley *et al.*, 1966; Glisin *et al.*, 1966), but because of the complications of hybridization techniques these experiments can only be considered as models; firm conclusions are not possible from the data so far available. Answering such simple questions as the number of active genes at various stages is just beginning to appear possible (Davidson and Hough, 1971; Gelderman *et al.*, 1971). In the case of sea urchins we do not know the basis for the very high rate of synthesis of heterogenous RNA at early stages. Is a larger percentage of the genome active in each cell during this time or are the individual genes just being transcribed more rapidly? If the number of active genes per cell decreases at gastrulation, it is not clear that the number of genes active in the total organism decreases since differentiation is producing more kinds of cells and each could have its own set of active genes.

In viewing the whole problem of the activation of specific genes at specific stages of development, one must be encouraged by the series of experiments on the function of the genes during cleavage (Nemer and Lindsay, 1969; Kedes and Gross, 1969; Kedes and Birnstiel, 1971). In this series a specific developmental function of producing more nuclei has been identified and the molecules involved, and their basic metabolism has been outlined. Although there are many basic questions which must still be answered in this one example, one may hope that we will be in a position in the next few years to add other such descriptive studies of the regulation of the synthesis of specific gene products to the bare bones of the gross description of RNA metabolism now available. One may look further into the future when the molecular control mechanisms regulating these sequences of synthesis can be elucidated.

III. Protein Synthesis

Examination of protein synthesis during early development has centered around two issues: the acceleration of protein synthesis at fertilization and the initial time of appearance of new proteins. The change in protein synthesis at fertilization appears to be due to activation of messenger RNA stored in the egg. It has been a model for studying the control of protein synthesis at the level of translation and has received wide interest. The appearance of new and/or specialized cell proteins is the most important basis for developmental progress and, therefore, central to the analysis of development.

A. ACCELERATION OF PROTEIN SYNTHESIS AT FERTILIZATION

1. Sea Urchins

The incorporation of exogenous, isotopically labeled amino acids into newly synthesized proteins increases greatly upon fertilization (Hultin, 1952). This increased incorporation is not prevented when RNA synthesis is stopped with actinomycin D, suggesting that more messenger RNA does not have to be synthesized for increased incorporation (Gross et al., 1964). Analysis of the increased incorporation showed that a large part of it was due to an increased permeability of the cells to exogenous amino acids but that some increase in actual protein synthesis also occurred (Epel, 1967). Although much previous work had suggested otherwise (for example, Malkin et al., 1964), Epel's results showed that significant protein synthesis was occurring in eggs before fertilization. My own calculations based on published data (Humphreys, 1969, 1971) suggest that the egg is synthesizing five and twenty times as much protein as an average cell in sea urchin plutei. Thus, on a per nucleus basis, the egg is synthesizing protein at a reasonably high rate. At fertilization this is increased to an even higher rate. Since the extra protein synthesis does not seem to require synthesis of more RNA, the messenger RNA necessary for the increase must be present in the egg before fertilization. I have shown that the efficiency of translation, the amount of protein made per messenger RNA molecule per unit time, does not change at fertilization (Humphreys, 1969). Neither the rate of attachment of ribosomes nor the rate of translocation of ribosomes along the messenger RNA molecule increases at fertilization. The increased protein synthesis thus cannot be dependent on the messenger RNA active in the egg but must be due to the translation of messenger RNA not previously being translated. The amount of messenger RNA activated at fertilization was calculated to be 6×10^{-12} gm/egg or about 0.4% of the total RNA in the egg (Humphreys, 1971).

There has been an extensive series of experiments from a number of different laboratories which have attempted to search for and identify the stored messenger RNA (Gross et al., 1965; Slater and Spiegelman, 1966) and/or to elucidate the mechanism of its activation (Hultin, 1964; Glisin and Glisin, 1964; Monroy et al., 1965; Verhey et al., 1965; Candelas and Iverson, 1966; Spirin and Nemer, 1965; Spirin, 1966; Mano and Nagano, 1966; Ceccarini et al., 1967; Piatigorsky, 1968; Metafora et al., 1971). Although these experiments have been direct, ingenious, and often pioneering, they have not led to a concensus concerning the nature of the stored messenger RNA or the mechanism of the translational control event. Indeed, in trying to summarize these, I find it difficult to use the conflicting data to predict the most likely possibilities or point to the most fruitful approaches for further experiments. Reasoning from my own limited data (Humphreys, 1969, 1971), I have pointed out that the regulation of translation must be achieved by controlling the activity of the messenger RNA molecules individually and cannot be achieved through some secondary component of the protein synthetic apparatus. This conclusion is based on the observation that each messenger RNA molecule becomes fully active and loaded with ribosomes within minutes after translation begins, even though the increase in active messenger RNA and protein synthesis occurs over a 2-hr period after fertilization (Humphreys, 1971). Thus I have argued that the translational control mechanism acts directly on the messenger RNA molecule and cannot act indirectly through ribosomes, transfer RNA, or other components. Such direct action may involve: (1) a change in the location of the messenger RNA in the cell so that it is available; (2) a change in the structure of the RNA so that it becomes translatable; or (3) an alteration in the association of the RNA with repressor or activator molecules. Unfortunately these three general possibilities include a broad spectrum of specific hypotheses for translational control and do not focus further experimental approaches in one, or even a few, directions.

I believe the most pressing problem at this juncture is to locate and identify the 0.4% of the RNA in the egg which is the stored messenger

RNA. This approach requires a suitable assay for the messenger RNA. A hybridization assay (Crippa and Gross, 1969) might be useful, but many technical problems still exist in such an assay (Gelderman et al., 1971). Other than this defined question, one is left with the approach of selecting a specific hypothetical mechanism on the basis of intuition or enthusiasm and seeking results consistent with it. Although it has been tried often in this system, this may still be the most suitable approach. To summarize, the acceleration of protein synthesis at fertilization in sea urchins has been shown to result from a translational level control of the activity of individual messenger RNA molecules. Analysis of this system has yet to focus on a specific molecular control mechanism.

2. Amphibians and Other Embryos

Although protein synthesis may change slightly on fertilization in a number of other embryos, most others do not show the dramatic change found in sea urchins. In amphibians there is no change in protein synthesis on fertilization; however, protein synthesis appears to increase rapidly during ovulation (Ecker and Smith, 1968). This increase occurs in the absence of a nucleus (Smith and Ecker, 1969) suggesting that the protein synthesis depends on messenger RNA stored in the egg. The activation of protein synthesis at this time in amphibians may be developmentally equivalent to the activation of protein synthesis at fertilization in sea urchins.

B. PROTEINS DURING EARLY DEVELOPMENT

Ideally, studies on the developmental regulation of synthesis of individual protein molecules would include a rigorous characterization of the protein molecule as a unique polypeptide chain, determination of its developmental functions, description of the timing and period of its synthesis, measurement of its absolute rate of synthesis, and elucidation of the regulatory mechanism which controls this synthesis. Needless to say, few studies have approached these goals. The most extensive studies deal with the synthesis of induced enzymes or specialized proteins in individual differentiating tissues and are beyond the scope of this review. The analysis of protein synthesis in early embryos is not well advanced but is very interesting because it occurs in a unique context.

1. Proteins in the Egg

Many kinds of experimental studies indicate that the egg is programmed to proceed almost automatically through early development until gastrulation. This programming is thought to include the accumulation during oogenesis of a large number of proteins which function to carry the embryo through early stages. In addition, some proteins are synthesized during early stages on the messenger RNA which was accumulated in the egg and activated at fertilization. The synthesis of the proteins which depend on nuclear function for the synthesis of their messenger RNA does not predominate until gastrulation. Unfortunately, very few examples of specific proteins are available to illustrate these generalizations. Although the generalization must have a certain validity, I am sure our present insight into the role of preformed proteins, proteins synthesized on preformed messenger RNA, and proteins dependent on nuclear function do not allow us to judge the importance of this generalization. Indeed, I would not be overly surprised if it turned out to be trivial.

Probably the most exciting suggestions of preformed protein components are those showing localized cytoplasmic elements which determine the fate of the cells cleaved from that area. For example, differentiation of sex cells seems to be controlled by localized materials in many eggs (Blackler, 1958). However, it is impossible to estimate the number of tissues specified by such elements. There may be only a handful in regulative eggs, a few more in determinative eggs, or there could be many in all kinds of eggs. Spindle proteins are accumulated in considerable quantities during oogenesis (Raff *et al.*, 1971) and must function during cleavage. But even here it has not been established that these proteins are accumulated much more extensively in the egg than in the cytoplasm of any other cell type. There is little evidence to identify other proteins which might be specially accumulated to carry the egg through early development. Except for the yolk the large egg may only provide general cytoplasm having ribosomes, mitochondria, etc., but otherwise be rather unspecialized. Only further analysis of egg components will elucidate the role egg proteins play in subsequent development.

2. *Proteins from Preformed Messenger RNA*

The proteins synthesized on the messenger RNA stored in the egg are also not well defined. So far it has not been possible to resolve them into individual peaks containing single species of protein molecules by either acrylamide gel electrophoresis or column chromatography (Spiegel *et al.*, 1965; Ellis, 1966; Gross, 1966; Mackintosh and Bell, 1969; Terman, 1970). This failure indicates that there are probably at least 50 to 100 different proteins and possibly many more. Hybridization studies show that the RNA available in the egg could code for thousands of proteins (Davidson and Hough, 1971) but there is no evidence how much of this RNA is actually translated. The function of these many proteins is not known. Inhibition of protein synthesis stops cleavage (Hultin, 1961; Hogan and Gross, 1971), suggesting that these proteins function in cleavage. They may be required for DNA synthesis since it stops when protein synthesis is inhibited (Young *et al.*, 1969). Some rather difficult and well-executed experiments have shown that a small percentage of the proteins synthesized on preformed messenger RNA resemble spindle or microtubular protein (Raff *et al.*, 1971, 1972). The egg has a large supply of these proteins and synthesis of slightly more on the stored messenger RNA cannot be of much significance unless they are a special class of proteins. At the moment the various possibilities cannot be further evaluated.

It has been suggested that the change in rate of protein synthesis at fertilization does not involve the synthesis of new kinds of proteins but may only be a more rapid continuation of the protein synthesis occurring in the egg (Mackintosh and Bell, 1969). Other studies have suggested that there is a sequential activation of stored messenger for different proteins (Terman, 1970). These suggestions raise important questions but are based on results with technical limitations and cannot be considered firm conclusions. However, it is within present capabilities to answer these questions. Once we can isolate proteins with 100% yields and define them as individual molecules, for example, as single bands of a purified radioactive protein on acrylamide gels (Raff *et al.*, 1971), it would then be possible to measure the absolute change in the rate of synthesis of these individual proteins by using the incorporation of radioactivity during labeling with saturating concentrations of radioactive amino acids (Berg, 1970).

No good generalization exists that explains why there should be protein synthesis upon stored messenger RNA. Since the amount of messenger RNA appears to be large compared to that usually synthesized by a single nucleus, one might postulate that these proteins are needed in quantities larger than is possible to obtain quickly from the activity of a single or a few nuclei. However, we know neither how much of any one kind of messenger RNA is present nor why the messenger RNA rather than the proteins is accumulated. It will not be possible to answer such questions until the details of activation of protein synthesis, the general nature of the proteins synthesized on the stored messenger RNA, and their role in development have been elucidated.

3. *Proteins Dependent on Nuclear Activity*

From the time that the major cell types are determined at gastrulation and onward through development, formation of the proteins necessary for development of the various tissues must be mostly dependent on nuclear activity occurring at about the same time. The many issues involved in the regulation of protein synthesis during the development and specialization of each tissue is beyond the scope of this review. The point of interest in the context of early development is the initiation of protein synthesis dependent on nuclear activity after fertilization and the nature of the proteins so synthesized.

Many lines of experimental evidence have suggested that early protein synthesis occurred on preformed messenger RNA and only later did newly synthesized messenger RNA play an important role (see Davidson, 1968, for full discussion of these ideas). Attempts to define the time of first synthesis of some individual protein or enzyme after fertilization have generally shown that it was not synthesized during early stages. Although in many cases this is a reasonable result in terms of developmental history, most such studies have not paid attention to defining rates of synthesis on a per nucleus basis and accounting for the large amount of protein synthesis occurring on preformed messenger RNA at the earliest stages. For example, one would not expect the synthesis of a specialized protein such as collagen until differentiation of the tissues which accumulate large amounts of collagen (Green et al., 1968). However, the increasing ease with which collagen is detected during development cannot logically be equated with increased rates of synthesis during development because more nuclei are being formed and background protein synthesis not dependent on concurrent nuclear activity is dropping as development proceeds. Studies attempting to determine the time at which nuclear function begins by determining the time of appearance of a paternal protein (for a recent example, Chapman et al., 1971) are also affected by such considerations. Recent discoveries that some proteins dependent on nuclear activity are synthesized very actively during the early cleavage stages (Nemer and Lindsay, 1969; Kedes and Gross, 1969) show that the result depends on the protein selected for study.

The accumulation of radioactive RNA in a special class of small polysomes soon after fertilization provided the first indication that RNA synthesized right after fertilization might be involved in protein synthesis (Spirin and Nemer, 1965; Infante and Nemer, 1967). At the time, the observers were probably much influenced by the idea that immediate nuclear activity was not necessary for early development and thought that these polysomes were inactive in protein synthesis. However, following some observations that early histone synthesis was very active (Lindsay, 1969), it was suggested that these small polysomes dependent on new messenger RNA synthesis were responsible for the synthesis of histonelike proteins (Nemer and Lindsay, 1969). The appearance of the messenger RNA in the small polysomes stops when DNA synthesis is inhibited suggesting that this protein synthesis is associated with the S phase of the cell cycle (Kedes and Gross, 1969). Data from my experiments (Humphreys, 1971) show that the first entry of RNA into small polysomes and presumably the first protein synthesis occur between 30 and 60 min of development when the first round of DNA synthesis occurs (Hinegardner *et al.*, 1964). Although there has been some argument that these proteins are the usual histones, the best data suggests that histones are synthesized on stored messenger RNA (Lindsay, 1969; Johnson and Hnilica, 1971). The results from sea urchins may also be true of amphibians since histonelike proteins seem to be the first to show qualitative changes during amphibian development (Spiegel *et al.*, 1970). These experiments show that protein synthesis associated with the formation of cleavage nuclei is dependent on nuclear activity after fertilization. It is not yet clear what nuclear proteins are actually involved or if this is the only class of proteins dependent on new RNA synthesis at these early stages.

It is surprising to me that nuclear proteins should be so synthesized during this critical period of development when specification of basic cell differentiations are occurring. One might expect such proteins to be localized in the egg and, indeed, the important ones might be. Likewise, I see no logical reason to predict that any other class of proteins should be included in this group. The initiation of nuclear activity and the role it plays in early protein synthesis and normal development should prove to be a very interesting subject in the coming years. Because of the emerging data I would not be surprised to see it become evident that they play a much more important role in the very first stages of development than has previously been thought.

C. SUMMARY

Early development provides a very interesting context for examination of protein synthesis. The egg is a large cell to provide the cytoplasm for the many cells produced during cleavage and may also contain many special proteins which function in cleavage and differentiation. In addition to these proteins synthesized during oogensis, new proteins required for the progress of development are synthesized beginning immediately after fertilization. Some of these proteins are synthesized on messenger RNA accumulated in the egg in an inactive state while others are synthesized on messenger RNA produced after fertilization. For the most part the proteins in all three categories and their roles in development have not been characterized and as yet no general principle for the inclusion of a particular kind of protein in one or another class has been developed.

Acknowledgments

The work from this laboratory was supported by grants Nos. 03480 and 06574 from the National Institutes of Child Health and Human Development and were performed with the expert assistance of Ronald Van Boxtel.

References

Aronson, A. I. (1972). *Nature (London) New Biol.* **235**, 40.
Belitsina, N. V., Aitkhozhin, M. A., Garrilova, L. P., and Spirin, A. S. (1964). *Biokhimiya* **29**, 363.
Berg, W. (1970). *Exp. Cell Res.* **60**, 210.
Blackler, A. W. (1958). *J. Embryol. Exp. Morphol.* **6**, 491.
Brandhorst, B. P. (1971). Ph.D. Thesis, University of California, San Diego.
Brandhorst, B. P., and Humphreys, T. (1971). *Biochemistry* **10**, 877.
Brandhorst, B. P., and Humphreys, T. (1972). *J. Cell Biol.* **53**, 474.
Brown, D. D., and Caston, J. D. (1962). *Develop. Biol.* **5**, 412.
Brown, D. D., and Littna, E. (1964). *J. Mol. Biol.* **8**, 669.
Brown, D. D., and Littna, E. (1966). *J. Mol. Biol.* **20**, 81.
Candelas, G. C., and Iverson, R. M. (1966). *Biochem. Biophys. Res. Commun.* **24**, 867.
Ceccarini, C., Maggio, R., and Barbata, G. (1967). *Proc. Nat. Acad. Sci. U.S.* **58**, 2235.
Chamberlain, J. (1967). *Biol. Bull.* **133**, 461.
Chamberlain, J. (1970). *Biochim. Biophys. Acta* **213**, 183.
Chapman, V. M., Whitten, W. K., and Ruddle, F. H. (1971). *Develop. Biol.* **26**, 153.
Collier, J. R. (1966). *Curr. Top. Develop. Biol.* **1**, 39.
Comb, D. G., Katz, S., Branda, R., and Pinzino, C. (1965). *J. Mol. Biol.* **14**, 195.
Craig, S. P. (1970). *J. Mol. Biol.* **47**, 615.
Crippa, M., and Gross, P. R. (1969). *Proc. Nat. Acad. Sci. U.S.* **62**, 120.
Davidson, E. H. (1968). "Gene Activity in Early Development." Academic Press, New York.
Davidson, E. H., and Hough, B. R., (1971). *J. Mol. Biol.* **56**, 491.
Davidson, E. H., Allfrey, V. G., and Mirsky, A. E. (1964). *Proc. Nat. Acad. Sci. U.S.* **52**, 501.
Dawid, I. (1966). *Proc. Nat. Acad. Sci. U.S.* **56**, 269.
Denis, H. (1966). *J. Mol. Biol.* **22**, 285.
Ecker, R. E., and Smith, L. D. (1968). *Develop. Biol.* **18**, 232.
Ellem, K. A. D., and Gwatkin, R. B. L. (1968). *Develop. Biol.* **18**, 311.
Ellis, C. H. (1966). *J. Exp. Zool.* **163**, 1.
Emerson, C. P., and Humphreys, T. (1970). *Develop. Biol.* **23**, 86.

Emerson, C. P., and Humphreys, T. (1971). *Anal. Biochem.* **40**, 254.

Emerson, C. P., and Humphreys, T. (1971a). *Science* **171**, 898.

Epel, D. (1967). *Proc. Nat. Acad. Sci. U.S.* **57**, 899.

Gelderman, A. H., Rake, A. V., and Britten, R. J. (1971). *Proc. Nat. Acad. Sci. U.S.* **68**, 172.

Giudice, G., and Mutolo, V. (1967). *Biochim. Biophys. Acta* **138**, 276.

Glisin, V. R., and Glisin, M. V. (1964). *Proc. Nat. Acad. Sci. U.S.* **52**, 1548.

Glisin, V. R., Glisin, M. V., and Doty, P. (1966). *Proc. Nat. Acad. Sci. U.S* **56**, 285.

Gould, M. (1969a). *Develop. Biol.* **19**, 460.

Gould, M. (1969b). *Develop. Biol.* **19**, 482.

Green, H., Goldberg, B., Schwartz, M., and Brown, D. D. (1968). *Develop. Biol.* **18**, 391.

Gross, P. R. (1966). *Curr. Top. Develop. Biol.* **2**, 1.

Gross, P. R., Malkin, L. I., and Moyer, W. A. (1964). *Proc. Nat. Acad. Sci. U.S.* **51**, 407.

Gross, P. R., Malkin, L. I., and Hubbard, M. (1965). *J. Mol. Biol.* **13**, 463.

Harris, S. E., and Forrest, H. S. (1967). *Science* **156**, 1613.

Hartmann, J. F., and Comb, D. G. (1969). *J. Mol. Biol.* **41**, 155.

Hinegardner, R. T. (1969). *Biol. Bull.* **137**, 465.

Hinegardner, R. T., Rao, B., and Feldman, D. E. (1964). *Exp. Cell Res.* **36**, 53.

Hogan, B., and Gross, P. R. (1971). *J. Cell Biol.* **49**, 692.

Hultin, T. (1952). *Exp. Cell Res.* **3**, 494.

Hultin, T. (1961). *Experientia* **17**, 410.

Hultin, T. (1964). *Develop. Biol.* **10**, 305.

Humphreys, T. (1969). *Develop. Biol.* **20**, 435.

Humphreys, T. (1971). *Develop. Biol.* **26**, 201.

Infante, A. A., and Nemer, M. (1967). *Proc. Nat. Acad. Sci. U.S.* **58**, 681.

Johnson, A. W., and Hnilica, L. S. (1971). *Biochim. Biophys. Acta* **246**, 141.

Kedes, L. H., and Birnstiel, M. L. (1971). *Nature (London), New Biol.* **230**, 165.

Kedes, L. H., and Gross, P. R. (1969). *Nature (London)* **223**, 1335.

Kijima, S., and Wilt, F. H. (1969). *J. Mol. Biol.* **40**, 235.

Lindsay, D. T., (1969). *Ann. Embryol. Morphog.* **1**, Suppl., 277.

MacKintosh, F. R., and Bell, E. (1969). *Science* **164**, 961.

Malkin, L. I., Gross, P. R., and Romanoff, P. (1964). *Develop. Biol.* **10**, 378.

Mano, Y., and Nagano, H. (1966). *Biochem. Biophys. Res. Commun.* **25**, 210.

Markert, C. L., and Cowden, R. R. (1965). *J. Exp. Zool.* **160**, 37.

Metafora, S., Felicetti, L., and Gambino, R. (1971). *Proc. Nat. Acad. Sci. U.S.* **68**, 600.

Monroy, A., Maggio, R., and Rinaldi, A. M. (1965). *Proc. Nat. Acad. Sci. U.S.* **54**, 107.

Nemer, M. (1963). *Proc. Nat. Acad. Sci. U.S.* **50**, 217.

Nemer, M., and Infante, A. A. (1967). *J. Mol. Biol.* **27**, 73.

Nemer, M., and Lindsay, D. T., (1969). *Biochem. Biophys. Res. Commun.* **35**, 156.

Patterson, J. B., and Stafford, D. W. (1971). *Biochemistry* **10**, 2775.

Piatigorsky, J. (1968). *Biochim. Biophys. Acta* **166**, 142.

Piatigorsky, J., and Whiteley, A. H. (1965). *Biochim. Biophys. Acta* **108**, 404.

Piko, L. (1970). *Develop. Biol.* **21**, 257.

Piko, L., Tyler, A., and Vinograd, J. (1967). *Biol. Bull.* **132**, 68.

Raff, R. A., Greenhouse, G., Gross, K. W., and Gross, P. R. (1971). *J. Cell Biol.*
  50, 516.
Raff, R. A., Colot, H. V., Selvig, S. E., and Gross, P. R., (1972). *Nature (London)*
  235, 211.
Roeder, R. G., and Rutter, W. J. (1969). *Nature (London)* **224,** 234.
Sconzo, G., Bono, A., Albanese, I., and Giudice, G. (1972). *Exp. Cell Res.* **72,** 95.
Slater, D. W., and Spiegelman, S. (1966). *Proc. Nat. Acad. Sci. U.S.* **56,** 164.
Smith, L. D., and Ecker, R. E. (1969). *Develop. Biol.* **19,** 281.
Smith, L. D., and Ecker, R. E. (1970). *Curr. Top. Develop. Biol.* **5,** 1.
Spiegel, M., Ozaki, H., and Tyler, A. (1965). *Biochem. Biophys. Res. Commun.*
  21, 135.
Spiegel, M., Spiegel, E. S., and Meltzer, P. S. (1970). *Develop. Biol.* **21,** 73.
Spirin, A. S. (1966). *Curr. Top. Develop. Biol.* **1,** 1.
Spirin, A. S., and Nemer, M. (1965). *Science* **150,** 214.
Terman, S. A. (1970). *Proc. Nat. Acad. Sci. U.S.* **65,** 985.
Verhey, C. A., Moyer, F. H., and Iverson, R. M. (1965). *Amer. Zool.* **5,** 637.
Whiteley, A. H., McCarthy, B. J., and Whiteley, H. R. (1966). *Proc. Nat. Acad.
  Sci. U.S.* **55,** 519.
Woodland, H. R., and Graham, C. F. (1969). *Nature (London)* **221,** 327.
Young, C. W., Hendler, F. J., and Karnofsky, D. A. (1969). *Exp. Cell Res.* **58,**
  15.

2

Developmental Regulation in Cotton Seed Embryogenesis and Germination

LEON S. DURE, III

I. Introduction

For the past six years we have been investigating the developmental biochemistry of seed embryogenesis and germination. Our overall goal has been an increased comprehension of the manner by which the ordered, sequential, and preprogrammed release of information from an organism's genome is controlled so as to give rise to the specialization of tissues in structure and function, constituting development and maturation. Initially, such a study involves a temporal mapping of the acquisition of structures and enzymatic capabilities by the developing tissue. Ultimately, such a study involves coming to grips with mechanisms which cause these capabilities to show up in the proper sequence and at the proper time so as to result in an integrated organism.

It is axiomatic that the orderly acquisition of capabilities and specialized functions is inherent in the organism's DNA, which means that the regulation of differential gene expression is somehow contained in the total genetic information. Further, it is axiomatic that these capabilities and specializations are manifested in cells by the synthesis of new and varied proteins via the informational intermediary, RNA. What are not obvious are the mechanisms which allow only discrete portions of genetic information to be manifested with time and which express different portions of the total genetic information in different cell types.

The embryogenesis and subsequent germination of plant seeds offer excellent systems for investigating regulatory mechanisms that operate to bring about cell differentiation and maturation. In many respects all plant seeds progress from the zygote to the young plantlet in a similar fashion. In embryogenesis, there is the creation of similar tissue types and the storing of nutritional reserves. In germination, mobilization of these reserves takes place to sustain the growth required to get the young plantlet in position to carry out photosynthesis and thus begin autotrophic growth. However, in the evolution of seed-bearing plants these processes of embryogenesis and germination have undergone extensive modifications. These changes have provided a number of variations on the common theme of how the regulation of gene expression brings about the progression from zygote to young plantlet. With the hope of correlating and integrating information obtained from different experimental approaches, we have concentrated our studies on a single tissue, the cotyledon tissue, of the embryo of the cotton plant (*Gossypium hirsutum*).

II. Cotton Cotyledons as a Developmental System

Figure 1 gives a schematic presentation of some of the gross aspects of the development of this tissue during embryogenesis and germination along a temporal axis. Some of the information presented in this figure was obtained from experiments which will subsequently be discussed in some detail. Figure 1 shows that during the first 3 weeks after pollen release (anthesis) the cotyledons increase in wet weight at a rather trivial rate. (However, during this period the endosperm tissue proliferates rapidly and attains its maximum size.) After this period the cotyledons begin a logarithmic growth phase that lasts for about 2 weeks and which is followed by a slower growth phase until their final size is reached about day 50. The dry weight increase follows the same general pattern. The growth of the cotyledon during embryogenesis is supported by the endosperm tissue which is totally consumed by the developing embryo during this period. This process of endosperm absorption by the embryo not only provides the embryo with the constitutents for building new cells but also provides for the deposition in the cotyledon cells of many large protein granules (aleurone grains) and numerous lipid droplets. These nutritional reserves are utilized to sustain the growth and development of the embryo during early germination.

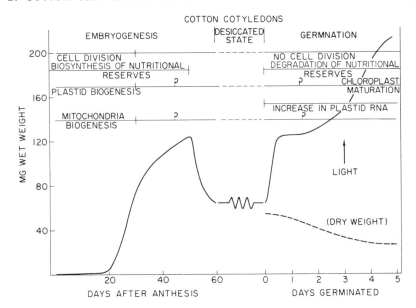

FIG. 1. Schematic presentation of cotton cotyledon development.

This total incorporation of endosperm nutrition by the embryo and its redeposition as protein and lipid in cotyledon cells is one extreme of the endosperm–embryo relationship in embryogenesis that is characteristic of many familiar plants such as beans and peas. Monocotyledonous plants (such as cereals and grasses) represent the other extreme in which endosperm absorption by the embryo does not take place until germination. Intermediate between these extremes are plants such as the castor bean in which some, but not all, of the endosperm is absorbed by the embryo and deposited as nutritional reserves in the cotyledon cells during embryogenesis.

In cotton, embryogenesis is halted by the drying up and hardening (sclerification) of the seed coat wall encompassing the embryo. The embryo desiccates during this period, and the final product of this process is the mature dry seed.

The water lost during the desiccation period is rapidly regained through imbibition during the first few hours of germination, but unlike embryogenesis, the subsequent massive increase in wet weight of the cotyledons caused by cell enlargement is not accompanied by an increase in dry weight, but rather is accompanied by a dry weight loss as the stored nutrients are mobilized and transported to the rapidly growing shoot and root axis. By the third day of germination the cotyledons have emerged from the soil and are in position to receive the light stimulus that brings about the final maturation of its chloroplasts.

Other aspects of Fig. 1 will be brought out in the presentation of our experimental results.

III. Nucleic Acid Levels during Cotyledon Development

One of our initial projects was to measure the levels of DNA, rRNA, and tRNA in the cotyledon tissue during the developmental sequence shown in Fig. 1 to determine if the preparation for any of the developmental steps involves an abrupt change in the number of ribosomes or tRNA molecules per cell (per unit of DNA). Our procedure for accomplishing this involved the rigorous extraction of total nucleic acid from cotyledon tissue in various stages of development beginning with cotyledon pairs whose wet weight was about 10 mg and ending with cotyledons from 5-day germinated seedlings. The solubilized nucleic acid was purified by conventional phenol–sodium dodecyl sulfate (SDS) methods followed by alcohol and cetyltrimethyl ammonium bromide precipitations. The total amount of nucleic acid realized from each preparation was determined by spectral analysis. Aliquots were then subjected to SDS-polyacrylamide gel electrophoresis (Loening, 1967), and the relative amounts of DNA, rRNA, 5 S RNA, and tRNA determined from spectrophotometric scans of these gels at 260 nm.

In Fig. 2 the results of these determinations are presented as μg nucleic acid/cotyledon pair vs. the wet weight of cotyledon pairs (for embryogenesis) or vs. days germinated. The lines given in Fig. 2 constitute the best line obtained for each nucleic acid type from over 50 individual preparations. The total solubilization of nucleic acids from this tissue is not routinely a reproducible operation, especially from cotyledons that contain a considerable amount of storage lipid. DNA solubilization in particular is often poor. Consequently, many determinations from lipid-rich cotyledons gave low DNA values per cotyledon pair. However, the majority of the samples gave values that fall within 10% of the values represented by the lines shown in Fig. 2. In samples that gave greater than 10% difference, the difference always represented lower values. The amounts of 5 S rRNA are included in the values for total rRNA. A detailed description of our procedures for isolating and estimating these nucleic acid values has been published (Ihle and Dure, 1972d).

Figure 2 shows that, in embryogenesis, all nucleic acid species accumulate in a linear fashion with an increase in cotyledon weight until the cotyledons reach 85 mg wet weight which corresponds to about three-

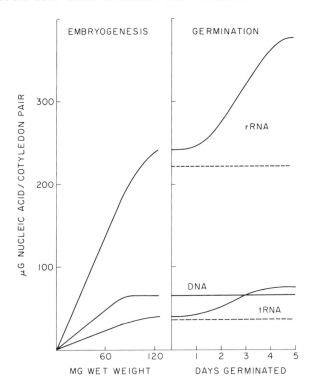

Fig. 2. Levels of nucleic acid species per pair of cotton cotyledons during embryogenesis and germination. Dashed lines represent levels of nonorganelle RNA species present during germination.

fifths their final size. At this point DNA accumulation ceases. This cessation, of course, implies that cell division stops in this tissue at this time as is pointed out in Fig. 1. Ribosomal and tRNA continue to accumulate after this point although at a somewhat slower rate relative to the increase in fresh weight. In view of the cessation of DNA accumulation, this continued increase in rRNA and tRNA represents an increase in ribosomes and tRNA molecules per cell. Thus the ratio of RNA to DNA changes during the last part of embryogenesis, yet the ratio of rRNA (including 5 S RNA) to tRNA remains constant throughout embryogenesis. Figure 2 also shows that when the embryos germinate, DNA synthesis in the cotyledons is not resumed. This indicates that the final number of cotyledon cells is reached at the point in embryogenesis corresponding to 85 mg wet weight. In addition to measuring DNA accumulation by determining the amount that can be extracted per cotyledon pair, we also ascertained qualitatively what nucleic acid species are synthesized by the cotyledons at various times in their development by incubating excised embryos and germinating seedlings in $^{32}PO_4{}^{3-}$ and measuring the amount of the isotope incorporated into the nucleic acids species which were separated by SDS-polyacrylamide gel electrophoresis. We were unable to detect any incorporation of the isotope into DNA after the 85-mg embryo stage in these experiments.

This finding that the DNA level in the cotyledons remains constant from the 85-mg stage through the fifth day of germination was to us very significant in that it indicated that cells that had been stockpiling large amounts of protein and lipid during embryogenesis reverse their metabolic direction and begin the degradation of these materials during germination without making new cells. This was significant since it made it likely that the biosynthesis of the enzymes that brought about this metabolic reversal and the induction of their synthesis could be studied without being obscured by the biosynthesis of all the necessary cell constitutents that accompanies cell division.

Further, Fig. 2 shows that rRNA and tRNA increase during germination. Here again, in the absence of DNA synthesis, this increase in RNA represents an increase in ribosomes and tRNA per cell. However, the spectrophotometric scans of the gels obtained with nucleic acid extracted during this germination period show that the increase in rRNA can be accounted for exclusively by an increase in chloroplastic rRNA. Figure 3 shows representative gel A_{260} profiles of nucleic acid from dry seed cotyledons and from 5-day germinated cotyledons. (These gels have been run for 5 hr to separate cytoplasmic and chloroplastic rRNA species, and consequently no tRNA or 5 S RNA remained on the gels.) A small but measurable amount of proplastid rRNA can be found in the dry seed cotyledon preparation that we estimate to represent about 8% of the total rRNA. This proplastid rRNA in the same relative amount is found in all nucleic acid preparations from embryogenic cotyledons of all ages suggesting that proplastid ribosomes are present throughout embryogenesis. The scan of the gel containing nucleic acid from the 5-day germinated cotyledons shows the increase in the plastid rRNA. The scan also shows the 23 S plastid rRNA to be largely degraded to smaller fragments. This degradation is routinely encountered with this plastid rRNA species from numerous different plants (Ingle, 1968), but it does not preclude an estimation of plastid rRNA which can be obtained by subtracting the cytoplasmic rRNA from the total rRNA. Many measurements of this sort lead us to feel that the cytoplasmic rRNA (and hence cytoplasmic ribosome number per cell) does not increase during the first 5 days of germination.

Since plastid tRNA cannot be distinguished from cytoplasmic tRNA

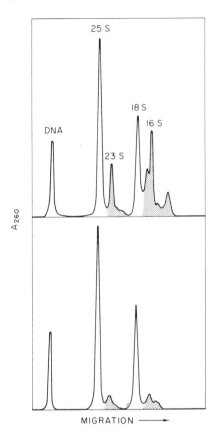

FIG. 3. Spectral scans of SDS gel electropherograms of total nucleic acid prepared from the cotyledons of dry seeds (lower scan) and from 5-day germinated cotyledons (upper scan).

in gel profiles, we have estimated the increase in plastid tRNA during germination from other experiments. In studies that will be subsequently presented here, we have been able to identify and measure the amounts of chloroplast isoaccepting species for several amino acids. These chloroplast species comprise about 8–10% of the total tRNA for these amino acids in embryogenic and dry seed cotyledons and about 50% of the total tRNA for these amino acids in 5-day germinated cotyledons. Assuming that the cytoplasmic isoaccepting species do not change in amount during germination, the increase in chloroplastic species from 8–10% to 50% of the total tRNA would account for the total increase in tRNA observed during germination. By extending the calculations made possible by the data given in Fig. 2, it would appear that the plastid rRNA and tRNA undergo a sevenfold increase per cell during the first 5 days of germination. This increase is observed in cotyledons that are maintained in the dark (etiolated) for the entire 5 days as well as those allowed to green after the third day. Furthermore, ascribing the increase in rRNA and tRNA observed during germination solely to plastid RNA shows that the ratio of tRNA to rRNA by weight is maintained at 1:6 for cytoplasmic RNA and 1:5 for plastid RNA, which are the ratios that apparently exist throughout embryogenesis. When the difference in molecular weight of cytoplasmic and plastid rRNA is then taken into consideration, these ratios indicate that there are between 13 and 14 tRNA molecules per ribosome for both cytoplasmic and plastid systems, and apparently this relationship is maintained throughout the development and maturation of cotyledons for both cytoplasmic and chloroplastic protein synthesizing systems. It is curious that this ratio is almost identical to that reported to exist in *E. coli* (Watson, 1970). Further, the data in Fig. 2 indicates that on a molar basis plastid ribosomes become nearly equal to cytoplasmic ribosomes in this tissue by the fifth day of germination as do plastid tRNA molecules relative to cytoplasmic tRNA molecules as shown by our tRNA studies. Ingle (1968) has observed a similar 1:1 molar ratio of chloroplastic to cytoplasmic rRNA in germinated lettuce cotyledons.

This increase in plastid rRNA is not accompanied by any noticeable increase in total DNA as we have pointed out. We do not as yet know if chloroplast DNA increases during this time. Chloroplast DNA may represent such a small amount relative to the nuclear DNA that its increase during this period would not be detected by these techniques. On the other hand, the increase in plastid rRNA and tRNA may represent an increase per plastid and not an increase in the number of plastids per cotyledon.

We have not mentioned any contribution to these nucleic acid values by mitochondrial nucleic acid because we have not observed any organelle tRNA species that do not appear to be plastid in origin and have failed to extract significant amounts of rRNA from partially purified mitochondria. We conclude that mitochondria nucleic acid must contribute a very minor amount of the total nucleic acid of the cotyledon at the developmental stages studied here.

IV. Isoaccepting tRNA Species in Cotyledon Development

Coincident with the experiments presented above on the nucleic acid content of developing cotyledons, we undertook the project of examining the levels of isoaccepting tRNA species from this tissue at various stages of development. At that time the notion was current that protein synthesis could conceivably be regulated both quantitatively and qualitatively by fluxes in the levels of isoaccepting tRNA species. Anderson (1969) had demonstrated *in vitro* that the rate of translation of a specific oligonucleotide mRNA was reduced if a single tRNA whose anticodon was required in the translation was in lower concentration than the other tRNA species. Further, the discovery that many tRNA molecules contained substituted purines (the cytokinins) which when applied to plant tissue influenced basic developmental events such as cell division and differentiation (reviewed by Skoog and Armstrong, 1970) reinforced the notion that tRNA somehow is involved in regulating development.

To explore the possibility that changing levels of tRNA species might modulate the rate of synthesis of specific groups of proteins and thereby exert a regulatory effect on cotyledon development, we prepared tRNA from the cotyledons of (1) young embryos that were below 85 mg and still undergoing cell division; (2) mature dry seeds; (3) 5-day germinated etiolated seedling; and (4) 5-day germinated greened seedlings and from isolated chloroplasts. From these preparations, we determined the levels of tRNA that would accept each of the 20 amino acids, and further, we determined the number and relative levels of the isoaccepting species for twelve individual amino acids. In order to compare the values obtained from these cotyledon tRNA preparations with the tRNA population of a distinctly different tissue, we prepared tRNA from the root tissue of cotton seedlings and made the same determinations as for the cotyledon tissue.

In any study that purports to ascertain differences in the relative amounts of components against a developmental background, it is an obligation of the investigator to establish (1) that all the material is extracted from the tissue and (2) that all the material is extracted and purified without losing the biological activity to be assayed. If biological activity is lost during purification, the investigator must establish how much is lost. With this information readers can then judge for themselves the meaningfulness of the differences or similarities reported.

As for the first obligation, we believe that the tRNA preparations used in this work represent all the tRNA in the tissues, since we have never obtained significantly greater amounts by absorbancy measurements of tRNA per cotyledon pair even when the tissues were extracted with a series of detergents and no precautions were made to maintain acceptor activity. In the course of establishing the procedures for obtaining this maximum extraction of tRNA, we found an absolute requirement for deoxycholate in the initial homogenization medium. If it is omitted,

75–80% of the tRNA is sedimented at 10,000 g from the initial homogenation. Regarding the second obligation, two assumptions had to be made. These were (1) that the average molecular weight of cotton tRNA molecules is 25,000 and that cotton tRNA at a concentration of 1 mg/ml (pH 7.2, in 0.01 M MgCl$_2$) gives an absorbance at 260 nm of 25. Based on these assumptions, all the preparations of tRNA used in this work had a total amino acid acceptance capability of about 90%. The 10% tRNA that could not be enzymatically aminoacylated was either damaged in purification or did not contain C–C–A 3' terminal nucleotides at the time of extraction. In either case we have assumed that the loss of acceptance was random and not selective for specific species of tRNA. The details of our extraction and purification procedures and a description of other measures used to characterize the tRNA preparations have been reported (Dure and Merrick, 1971; Merrick and Dure, 1971; Merrick and Dure, 1972).

Preparations of crude aminoacyl tRNA synthetases were made from cotyledons at the various stages of development, and negligible differences between them were noted in the *extent* of charging the different amino acids. However, the *rates* of charging were much greater in many cases with the synthetase preparations from 5-day germinated cotyledons (which presumably contain a higher concentration of chloroplast synthetases). Consequently, these enzyme preparations were used throughout.

The percentage acceptance of these preparations for each of the amino acids is presented in Table I. It can be seen from this table that there is almost no change in the relative levels of tRNA for each amino acid in the cotyledons of young embryos and those from dry seeds. In addition to being a comparison of two stages of development, this is also a comparison between dividing and nondividing cells as is pointed out in Figs. 1 and 2. Furthermore, this constancy in the distribution of amino acid acceptance is reflected even in the tRNA prepared from roots! There are two instances only in which there does appear to be a substantial difference in the amount of tRNA that will accept a specific amino acid. These are (1) a low level of asparagine acceptor in the cotyledons from young embryos and (2) a low level of serine acceptor in the root preparations. At the present time we do not know if these differences are artifacts of unknown origin or meaningful changes in levels of specific isoaccepting tRNA.

Table I does show that there are changes in the relative concentration of amino acid acceptor molecules that occur during germination. That is, the amount of tRNA for certain amino acids is significantly different in the preparations from green cotyledons from that obtained from the

TABLE I

PERCENTAGE OF TOTAL tRNA CHARGED WITH EACH AMINO ACID

Amino acid	Embryo cotyledons	Dry seed cotyledons	Root	Green cotyledons	Chloroplasts
Ala	5.1	5.1	5.0	4.8	4.7
Arg	9.0	8.7	8.7	9.3	9.5
Asn	1.4	2.5	2.5	1.3	2.1
Asp	6.8	6.5	6.6	6.1	5.0
Cys	0.8	0.7	0.8	0.7	0.9
Gln	0.2	0.2	0.2	0.2	0.2
Glu	2.0	2.1	2.3	2.2	3.0
Gly	10.0	10.1	10.3	9.7	8.6
His	3.4	3.3	3.3	3.7	3.6
Ile	3.2	3.2	3.3	3.3	4.4
Leu	10.0	9.8	10.2	11.2	11.1
Lys	5.2	5.3	5.5	3.6	4.0
Met	3.5	3.3	3.5	4.7	5.8
Phe	4.6	4.6	4.7	5.0	6.7
Pro	4.6	4.6	4.4	4.1	3.2
Ser	3.4	3.2	1.3	3.8	4.1
Thr	5.6	5.6	5.7	5.4	4.7
Trp	2.1	1.9	1.9	2.3	1.9
Tyr[a]	2.7	2.7	2.7	2.7	3.0
Val	9.0	8.9	9.0	8.8	6.6
	92.6	92.3	91.9	92.9	93.1

[a] Cotton enzyme unstable. Estimate from values obtained with *E. coli* synthetase.

dry seed cotyledons. We have found that tRNA from the 5-day germinated etiolated cotyledons is no different from that obtained from green cotyledons in any parameter examined, and consequently separate values for tRNA from etiolated cotyledons is not presented. In addition, Table I shows that the composition of tRNA prepared from isolated and partially purified chloroplasts is distinct from that of green or dry seed cotyledons.

It has been pointed out earlier (Fig. 2) that we believe that chloroplast tRNA increases sevenfold during germination and reaches a concentration per cell nearly equal to the concentration of cytoplasmic tRNA which does not increase per cell to any measurable extent during this period. Should this be true, a change in the distribution of the tRNA molecules accepting specific amino acids would be expected when tRNA from dry seed cotyledons and from 5-day green cotyledons is compared, since the increase in chloroplast tRNA species would influence the total tRNA population to a much greater extent.

Our basis for believing that chloroplastic tRNA increases sevenfold during this early stage of germination is the pattern of change in the levels of individual tRNA species that we observed during early germination when we determined the number and relative levels of the isoaccepting species for several amino acids.

To visualize the number and relative levels of isoaccepting tRNA species for an individual amino acid we employed two techniques. The first of these involved the chromatographic separation of the various isoaccepting species acylated with a radioactive amino acid on the RPC I column developed by Wiess and Kelmers (1967). This column separates whole molecules of acylated tRNA on the basis of small differences in charge and in reactivity of the tRNA with a nonmobile organic phase on the column matrix. Transfer RNA's differing only trivially from one another can be separated on this column. However, spurious peaks (that is, elution peaks of radioactive aminoacyl tRNA that do not represent a unique oligonucleotide composition) are encountered with this column from a variety of causes, and this fact prompted us to resort to a second technique for resolving the different isoaccepting tRNA species and determining their relative amounts.

This technique involved the digestion of tRNA acylated with a specific radioactive amino acid by ribonuclease T_1. This enzyme, which specifically cleaves polynucleotide chains at guanine residues leaving the G $3'PO_4$, produces only one radioactive oligonucleotide fragment from each aminoacyl tRNA—the one carrying the radioactive amino acid. The length of this oligonucleotide depends on the position of the first guanine in from the C–C–A terminus of the tRNA chain. If this digestion mixture is chromatographed on DEAE cellulose at pH 4.5 without urea, radioactive oligonucleotides that differ in their nucleotide length or composition can be separated with a NaCl gradient. The different radioactive peaks, of course, represent fragments generated from different isoaccepting species. There may be isoaccepting species that do not differ in the nucleotide composition of the amino acid-containing fragment produced by the T_1 digestion. The existence of these isoaccepting species could not be ascertained by this technique, since their radioactive oligonucleotides would be identical and elute together. However, the different species demonstrated by this technique are unequivocally unique species. This technique in combination with the RPC I chromatographic profiles usually allows the total number and relative levels of isoaccepting species to be visualized.

Utilizing these methods we determined the distribution of isoaccepting tRNA species for a number of amino acids in cotyledon tRNA prepared from the developmental stages listed above and from chloroplasts and roots. These studies failed to show any change in the number and levels of nonchloroplastic isoaccepting species in the cotyledons at any developmental stage. Furthermore, the root tRNA showed the same distribution of isoaccepting species for the amino acids examined in this fashion. However, these studies did reveal a number of developmental features concerning chloroplastic tRNA. Here again, no change in the number of isoaccepting species was observed for chloroplastic tRNA nor in their levels relative to each other during cotyledon development. On the other hand, these species were found to increase markedly during the first 5 days of germination relative to the cytoplasmic species. Figure 4 is a set of DEAE column elution profiles obtained with [^{14}C]methionyl tRNA digested by T_1 RNase, and this figure shows the pattern of change we have obtained for eight additional amino acids to date (Dure and Merrick, 1971; Merrick and Dure, 1971, 1972). In this figure we see that if the tRNA is obtained from young embryo or dry seed cotyledons or from roots, two [^{14}C]methionyl oligonucleotides initially elute in small amounts and a third fragment elutes later and comprises the bulk of the radioactivity. As has been explained, these radioactive fragments represent unique tRNA^{met} species. If the tRNA is obtained from green cotyledons the amounts of the first two species is considerably increased relative to the third, and if the tRNA is obtained from partially purified chloroplasts (prepared by a modification of the nonaqueous procedure of Stocking, 1959), the first two species now comprise the bulk of the [^{14}C]methionyl oligonucleotides. It is from profiles such as these that we have been able to identify as chloroplastic tRNA species those which increase greatly in amount during germination and which are concentrated in partially purified chloroplasts. This figure presents another elution profile that further substantiates our presumption that those species that increase in amount during germination and are concentrated in chloroplasts are chloroplastic species. This is shown in the bottom right-hand profile which is of [^{14}C]methionyl tRNA prepared from green cotyledons and subsequently incubated with N-10-formyltetrahydrofolate and a cotton cotyledon enzyme preparation before being digested with the T_1 RNase. One of the chloroplastic methionyl oligonucleotide fragments is seen to have acquired greater net negative charge causing it to elute at a higher salt concentration. This is what would be expected of an N-formylmethionyl oligonucleotide. The radioactive amino acid recovered from this elution peak was shown to be N-formylmethionine, which demonstrated that one of the chloroplastic tRNA^{met} species is a tRNA$_f^{met}$. Chloroplasts have been shown to utilize N-formylmethionine

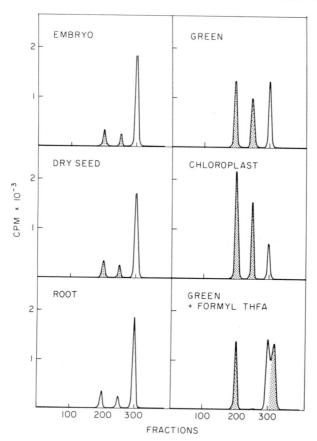

Fig. 4. Elution profiles of [¹⁴C]methionyl oligonucleotides from DEAE cellulose columns. Sources of tRNA are as indicated. CPM, Counts/min.

to initiate protein synthesis (Schartz *et al.*, 1967), whereas higher plant cytoplasmic protein synthesis has been shown to use nonformylated methionine (Leis and Keller, 1970).

Not shown in Fig. 4 is the profile obtained with tRNA from 5-day germinated etiolated cotyledons. This has been omitted since profiles from this tRNA are identical with those obtained from tRNA from green cotyledons, i.e., they show the same increase in chloroplastic species.

From profiles such as those presented in Fig. 4 the following points concerning chloroplastic tRNA can be inferred.

(*a*). It exists in low amounts in young embryo cotyledons, presumably being present even in the zygote.

(*b*). It exists in about the same low amounts in roots. This observation strengthens the proposition that root amyloplasts are derived from the same proplastid progenitor as are chloroplasts (Kirk and Tilney-Bassett, 1967).

(c). It increases about sevenfold per cell during the first 5 days of germination ultimately representing about one-half of the total cotyledon tRNA, and its increase can account for the total increase in the cotyledon tRNA observed during this phase of germination.

(d). Its increase does not require an induction by light.

We have tentatively concluded from these data at this point that, except for the low level of tRNA^{asn} in young embryo cotyledons, the maturation of cotton cotyledons is not accompanied by changes in cytoplasmic tRNA species either in number or relative amount. In fact, we are impressed with the apparent constancy in the levels of cytoplasmic tRNA species during this developmental sequence and with the near identity of the cotyledon cytoplasmic tRNA population to that of roots, a distinctly different tissue. Implicit in the idea that the synthesis of specific proteins might be modulated by changes in levels of tRNA species is the assumption that synonym code words do not occur randomly in mRNA, but have sorted themselves out so as to fit into a developmental scheme of regulation based on an increase or decrease in the rate of synthesis of specific proteins. If the genetic code is considered to be highly evolved toward the prevention of lethal mutations, then the idea that tRNA plays a direct role in the regulation of development appears, in retrospect, to have been a poor one.

These sorts of experiments, of course, do not measure any allosteric regulatory capabilities of charged or uncharged tRNA or of modified tRNA which may function to regulate the internal concentration of cell constituents (Silbert et al., 1966; Jacobson, 1971). However, they do tend to rule out a relationship between the population of cytoplasmic tRNA species and big developmental events that are outlined in Fig. 1.

V. Transcription without Translation in Embryogenesis

In addition to mapping temporally nucleic acid levels in developing cotyledons, we simultaneously undertook another approach to the problem of the control of sequential gene expression as manifested by cotyledon development. We decided to look for the appearance of new enzymes that is coeval with an abrupt developmental change, hoping that a study of their induction would prove fruitful.

We had earlier noted (Dure and Waters, 1965; Waters and Dure, 1966) that a great deal of the protein synthesis that occurs in the cotyledons during the first 3 days of germination (as determined by the incorporation in vivo of ^{14}C-labeled amino acids) is not sensitive to levels of actinomycin D that inhibit detectable RNA synthesis. This observation suggested that much of the protein synthesis that occurs during the first several days of germination is programmed by mRNA that exists in the dry seed, i.e., it is transcribed during embryogenesis. With this in mind, we decided to look for several enzymes that might be assumed to be necessary for germination but not essential for embryogenesis, to determine if these enzyme were synthesized *de novo* in germination, and, if this were true, to determine if their synthesis required concomitant RNA synthesis. We chose a proteolytic enzyme and isocitritase as likely candidates, since germination (but not embryogenesis) is characterized by the degradation of the storage protein and by the conversion of the stored lipid to carbohydrate in these cotyledon cells.

Crude extracts were made from cotyledons during their first 5 days of germination and assayed for isocitritase activity and for proteolytic activity (using the trypsin esterase substrate, benzoyl arginine ethyl ester). Both of these enzyme activities proved to be absent in dry seeds and in embryos, were found to show up after 24 hr of germination and to increase for several days hence, finally plateauing about the fifth day of germination. Furthermore, both of these activities arose normally in the presence of actinomycin D but were totally absent when seeds were germinated in the presence of cycloheximide. These findings suggested that these enzymes are synthesized *de novo* from mRNA existing in the dry seed.

In order to more precisely establish their *de novo* synthesis, a protocol for purifying the proteolytic enzyme to homogeneity was developed, and by utilizing this protocol the enzyme was purified from cotyledons that had been germinated in the presence of ^{14}C-labeled amino acids. The specific activity (counts/min/mg protein) of the purified protease proved to be 33 times higher than the crude Spinco supernatant protein from these cotyledons. This experiment more strongly indicates that the protease enzyme is synthesized *de novo* during germination. Subsequently, many of the properties of the purified protease were established and it was found to be a carboxypeptidase C type enzyme. (Ihle and Dure, 1972b,c).

Since the carboxypeptidase and isocitritase activities cannot be demonstrated in embryos nor in the dry seed prior to germination, the mRNA for these enzymes must somehow be restricted from being translated during the period between their transcription in embryogenesis and their use in germination. Fortunately, we were able to study this phenomenon of delayed translation because of the capability of the cotton embryo to precociously germinate when it is removed from the incipient seed coat tissues and placed on wet filter paper or agar gel or even in water and shaken. During this precocious germination, the cotyledons unfurl and expand as in normal germination and will green after several days in the light. The growth of the shoot and root are, likewise, no different from normal germinative growth except for being somewhat slower. Making use of precocious germination, we were able to demonstrate at what point the mRNA's for the carboxypeptidase and isocitritase are transscribed from their cistrons during embryogenesis. This was done by showing at what point in embryogenesis the appearance of these enzymes becomes actinomycin D-sensitive when the embryos are precociously germinated. This point turned out to correspond to the time in embryogenesis when the embryo wet weight is about 85 mg (three-fifths its final size) which is about 30 days after anthesis. Prior to this stage in embryogenesis neither enzyme activity develops in the presence of actinomycin D.

Although the isocitritase has not been purified to homogeneity and shown to be synthesized *de novo* in germination in the same manner that the carboxypeptidase was, its appearance during germination (precocious and normal) and its behavior with respect to inhibitors is precisely the same as the carboxypeptidase. Hence, we feel that it also is synthesized *de novo* in germination from preexisting mRNA that is transcribed at the 85-mg stage of embryogenesis. In fact, we consider that the carboxypeptidase and isocitritase represent an entire class of "germination" enzymes that are synthesized *de novo* during the first 5 days of germination from mRNA that is transcribed at this point in embryogenesis.

With the realization that the translation of the mRNA for the carboxypeptidase and isocitritase is delayed from the time of its synthesis until germination commences, it became obvious that we had a very nice example of one of the curious phenomena of developmental biochemistry. The separation in time of mRNA synthesis from its translation into protein was shown to be a developmental phenomenon in animal embryogenesis as early as 1963 (Gross and Cousineau). This phenomenon, in which the transcription of genes into mRNA occurs in one developmental stage, the mRNA "stored" in some as yet undelineated form, and subsequently is translated at a later stage, has now been observed in a large number of animal and fungal developmental studies. (reviewed by Gross, 1968; Davidson, 1969; Brown and Dawid, 1969.) Subsequent to the original report that protein synthesis in germinating cotton utilizes in part preformed mRNA (Dure and Waters, 1965), similar reports have appeared for germinating peanuts (Cherry, 1968), black-eyed peas (Chakravorty, 1968), and wheat (Chen *et al.*, 1968; Weeks and Marcus, 1971).

In two respects our system involving translation delay appeared to have an experimental advantage over many of the other systems.

First, we could assay two specific enzymes in studying the delay phenomena, and, second, we could apparently overcome the delay by dissecting the embryos from the maternal tissue and germinating them precociously. With this in mind, we began to look for mechanisms of translation control operating in this tissue that result in the prohibition of the translation of the mRNA for these germination enzymes during the last stages of embryogenesis.

The time course for the appearance of carboxypeptidase activity in cotyledons during the precocious germination of embryos larger than 85 mg is shown in Fig. 5. We found out that if these embryos were washed extensively in distilled water for a few minutes prior to placing them on agar gel to germinate precociously, the time course of carboxypeptidase appearance was much accelerated (Fig. 5A). Since removal of the embryo from its surrounding tissue (ovule tissue) is apparently all that is required to induce precocious germination and carboxypeptidase and isocitratase synthesis, and, since washing the embryo surface accelerated the synthesis of these enzymes, it appeared that the ovule tissue might be the source of a factor which is absorbed by the embryo and specifically prevents the translation of the mRNA for the germina-

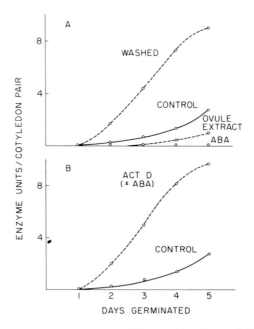

Fig. 5. The appearance of carboxypeptidase activity in cotyledon extracts during the precocious germination of embryos weighing 95 mg wet weight.

tion enzymes. The action of the factor in inhibiting translation would have to be selective since other protein synthesis does occur during this stage of embryogenesis. To test this idea we prepared an aqueous extract of the ovule tissue and tested its effect on the time course of the appearance of the two enzymes during precocious germination. Figure 5A shows that this extract delayed the appearance of the enzymes. The next experiment was to test the effect of abscisic acid (ABA) on the appearance of the two enzymes during precocious germination. This was an obvious experiment because this plant growth regulator is a potent inhibitor of many plant processes (reviewed by Addicott and Lyon, 1969) and was first isolated from cotton bolls (Okhuma *et al.*, 1963). Figure 5A shows that ABA totally inhibited the appearance of the enzymes. Subsequently we were able to isolate ABA from the aqueous extract of the ovule tissue obtained during the later stages of embryogenesis. Thus it appears that abscisic acid is the compound responsible for inhibiting the synthesis of the germination enzymes, presumably by being secreted by the ovular tissue and absorbed by the developing embryo. Such a mechanism appears feasible since the ovular tissue dies and sclerifies just before embryogenesis is complete, and hence is not a functioning tissue in germination.

The fact that dissected embryos do precociously germinate and synthesize the germination enzymes suggests that the ABA present in the embryo at the time of excision is gradually degraded by the embryo. However, a surprizing observation was made in the course of carrying out these experiments. It will be remembered that the time when the mRNA synthesis for the two germination enzymes commences was determined by precociously germinating dissected embryos in the presence of enough actinomycin D to stop further RNA synthesis. The two germination enzymes are synthesized in these embryos when precociously germinated in the presence of actinomycin D after this point. The observation was that not only are the enzymes synthesized in the presence of actinomycin D but that their synthesis was accelerated in time by actinomycin D to the same extent as that observed with washed embryos (Fig. 5B). Furthermore, when actinomycin D and ABA are supplied to the precociously germinating embryos simultaneously, the actinomycin D stimulus is still observed (Fig. 5B). This observation suggests that the inhibition of translation brought about by ABA requires concomitant RNA synthesis. In summary, all the results shown in Fig. 5 can be explained by assuming:

1. Dissected embryos larger than 85 mg have ABA on their surface that is produced by the ovule tissue and they are in the process

of absorbing it. Washing off this ABA accelerates the translation
of the mRNA for the germination enzymes.

2. Embryos contain a degradative system for ABA, since unwashed
   embryos develop the enzyme activities in time.
3. The inhibitory effect of ABA is somehow dependent on RNA synthesis, since actinomycin D accelerates enzyme synthesis even in
   the presence of ABA.

VI. Induction of Transcription in Embryogenesis

Since, at this point, we felt that we had found the time in embryogenesis when the transcription of the mRNA for the germination enzymes
begins and had implicated ABA as the primary effector in inhibiting
the translation of this mRNA during the remainder of embryogenesis,
we turned our attention to embryos smaller than 85 mg. *In vivo* cell
division is still going on in these younger embryos (see Figs. 1 and
2), but this abruptly stops on dissection. Consequently the final size
reached by the cotyledons during precocious germination becomes succes-
sively smaller as younger embryos are germinated, reflecting a succes-
sively smaller cell number. Carboxypeptidase and isocitritase activities
develop slowly when these embryos are precociously germinated, however, not in the presence of actinomycin D. Thus the appearance of
germination enzyme activities during the precocious germination of these
small embryos requires both transcription and translation. This transcription represents a premature induction, since it normally does not
occur until the 85-mg stage of development. Apparently no ABA is being
produced by the ovule wall at these earlier stages in embryogenesis,
since extracts of the ovule tissue have no inhibitory effect on the appearance of the enzymes in these small embryos, or in larger ones.

We were able to demonstrate that the transcription of the mRNA
for the germination enzymes begins within 24 hr after the dissection
of the embryos, since after 24 hr of precocious germination, actinomycin
D can be added to the germination matrix without inhibiting the appearance of the germination enzymes. It turned out that simply removing
the entire boll from the plant and waiting 24 hr before dissecting the
embryos from the boll is sufficient for inducing this premature transcription of this mRNA.

This observation suggested that the induction of this premature transcription results from breaking the vascular connection between the embryo and the mother plant. This is an attractive hypothesis since the
vascular connection with the mother plant normally breaks at the time

the funiculus connecting the seed coat with the placenta of the boll breaks which is about 30 days after anthesis when the embryo weighs about 85 mg. This, as has been shown, is when the mRNA for the germination enzymes is normally transcribed. Furthermore, it coincides with other events such as the appearance of ABA in the ovule tissue and the cessation of DNA synthesis and cell division in the embryo. Implicit in this hypothesis is the existence of a hormonal substance in the vascular solutes coming from the mother plant that maintain cell division in the embryo and prevent subsequent developmental events that take place after the 85-mg stage. From these data concerning the appearance of the carboxypeptidase and isocitritase during normal and precocious germination it is possible to erect a sequence of events along a temporal axis that takes place during the embryogenesis and germination of the cotyledons. Such a conjectural scheme is presented in Fig. 6.

This figure depicts vascular factor(s) entering the ovule tissue and embryo during early and mid-embryogenesis, maintaining DNA synthesis and cell division and somehow maintaining the cistrons for the germination mRNA in a repressed state. Rupturing this vascular flow stops cell division and induces the transcription of the mRNA and its subsequent translation, as has been shown.

About 30 days after anthesis, when the embryo weighs about 85 mg wet weight, this vascular connection is broken *in vivo* which brings about a number of basic developmental events. DNA synthesis stops as shown in Fig. 2 and the cistrons for the germination mRNA are derepressed. The germination enzymes will now appear in precocious germination in the presence of actinomycin D. The loss of the vascular

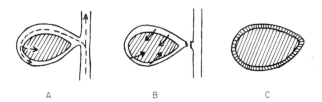

A   B   C

FIG. 6. A schematic presentation of developmental events in cotton cotyledon embryogenesis. (A) 20–30 Days after anthesis; vascular factors maintain DNA synthesis and cell division, the cistrons for germination are repressed, and excision induces transcription of germinaton enzymes. (B) 30–50 Days after anthesis; vascular connection is broken, DNA synthesis and cell division ceases, mRNA for germination enzymes is transcribed, and ABA production by ovule tissue begins which prevents precocious germination and the translation of the germination mRNA. (C) After 50 days after anthesis; ovule tissue is dead, removing source of ABA, desiccation of embryo begins, mRNA for germination enzymes is available for translation, and normal germination occurs upon inhibition.

factors also elicits the synthesis of ABA and its secretion by ovule tissue since extracts of this tissue from this stage on inhibit the appearance of the germination enzymes, whereas extracts from this tissue prior to this stage do not. The translation of the mRNA for the germination enzymes is proscribed during this period by the action of ABA which is absorbed by the cotyledons. The remaining embryo growth takes place at the expense of the ovular tissues.

As the embryo approaches its final size the ovule tissues begin to sclerify and subsequently die, thus removing the source of ABA. Finally the embryo desiccates to form the dry seed. This desiccation phase is not necessary for any further maturation since the germination of full-sized, but undesiccated, embryos is indistinguishable from that of dry seeds.

Oddly, ABA has no effect on the appearance of the carboxypeptidase or isocitritase in the cotyledons during the germination of mature seeds or of mature, undesiccated embryos. Whether this relates to a change in specific permeability or to a loss of receptor proteins in the cotyledons is not known. The details of this work on the regulation of the appearance of the germination enzymes can be found in the following references Ihle and Dure (1969, 1970, 1972a,d), and the characteristics and catalytic properties of the carboxypeptidase found in Ihle and Dure (1972b,c).

VII. Discussion and Prospect

The highlight, in our mind, of nucleic acid metabolism during the germination of cotton cotyledons is the enormous increase in chloroplast RNA which takes place against the rather quiescent background of nuclear inactivity in nondividing cells. Our measurement of chloroplastic rRNA and tRNA show that these components are maintained in the same low level per cotyledon cell during all the stages of embryogenesis, and that, based on our tRNA measurements, they are maintained at this low level in the growing root during germination. In this root tissue they are conceivably functioning in amyloplasts. This is an attractive idea in some respects. Both chloroplasts and amyloplasts seem to have all the requisite enzymes for synthesizing starch from carbohydrate intermediates, and one can conceive of an amyloplast as a chloroplast in which the genetic information (nuclear and/or organelle) for bringing into existence the light-harnessing apparatus remains repressed. This notion predicts that "chloroplast" DNA should be found in roots.

"Chloroplast" DNA is certainly present in the phloem parenchyma cells of carrot roots used by Steward (Steward *et al.*, 1958) in his classic demonstration of the totipotency of single cells.

Cotyledon germination, in contrast to root growth, is characterized by the sudden amplification of chloroplast RNA components which takes place in darkness as well as in light. This RNA synthesis presumably prepares the chloroplast for its last phase of maturation which is light induced. How plastid RNA synthesis is stimulated at this point in development in cotyledons presents a nice developmental problem.

One approach to this problem would be to test the possibility that chloroplastic RNA polymerase may be bacterialike and that its subunit composition may change in cotyledon cells when the time arrives for it to transcribe a new body of chloroplastic DNA. This idea is even more attractive when it is considered that the subunit(s) that confers specificity for different regions of the chloroplast genome may be coded for in the nuclear DNA. This would subordinate chloroplast development to nuclear control in a general sense, but at the same time allow many chloroplast enzymes and structural proteins to be encoded in chloroplastic DNA, and thus be maternally inherited.

The induction of the amplification of chloroplast RNA components that occurs in the dark and the light induction that triggers the final maturation of this organelle deserve to be more closely and sophisticatedly studied—preferably in tissues uncomplicated by concomitant cell division.

In the study of the transcription and translation of the mRNA for the germination enzymes, for which the carboxypeptidase and isocitritase serve as representatives, an important point remains to be established, namely, how long a time is required by the embryonic cotyledons to complete the transcription of the mRNA for the germination enzymes. We have yet to measure the number of carboxypeptidase molecules formed during precocious germination in actinomycin D in cotyledons that weigh, 85, 90, 95 mg, etc. These measurements should allow us to determine how long a transcriptional period is required to accumulate the mRNA that is normally present in the mature seed for these enzymes.

We feel very fortunate that this cotton cotyledon system has provided us with a tissue in which we can to some degree control the transcription and translation of mRNA for specific enzymes. As we have shown, premature transcription of these mRNA's can be induced by dissecting and precociously germinating embryos that weigh less than 85 mg. We can inhibit their subsequent translation with ABA, and finally bring about their translation with actinomycin D or by removing the ABA. This capability may be useful in tackling some of the basic questions that these experiments themselves have raised. Some of the obvious questions are:

1. What is the physical form in which the stored mRNA exists during late embryogenesis?
2. What is the mechanism of action by which abscisic acid inhibits the translation of the stored mRNA during late embryogenesis?
3. What is the factor(s) that presumably enter the embryo from the maternal tissue during early and mid-embryogenesis that maintain cell division and keep the cistrons for the germination mRNA's repressed?
4. How do these factors operate to control embryo growth during this period and what happens molecularly when the supply of these factors is cut off that results in the cessation of cell division, in the transcription of the germination mRNA's, and in the synthesis of abscisic acid by ovule tissues?

As a final consideration, a comparison should be made of our observations on cotton seed embryogenesis and germination with what is known about the biochemistry of germinating barley, because it may provide a molecular insight into the evolution of monocots from dicots. It is tempting to view barley (monocot) embryogenesis as coming to a halt at a point somewhat equivalent to the midpoint of cotton (dicot) embryogenesis. At this point in embryogenesis most of the nutrition for supporting germination growth remains as endosperm outside the embryo in both plant types. Also at this point the appearance of the degradative enzymes for mobilizing this nutrition still requires both transcription and translation. Cell division in cotton cotyledons stops close to this point just as the vestigal cotyledon of barley (the scutellum) experiences little or no cell division in germination. In the final 20 days of cotton embryogenesis, precocious germination is sensitive to ABA, whereas the normal germination of barley is antagonized by this growth regulator. In fact, one of the well-documented sites of ABA inhibition that occurs in barley germination is the inhibition of degradative enzyme synthesis (Chrispeels and Varner, 1967). Seen in this view, the early stages of monocot germination are analogous in many respects to the completion of dicot embryogenesis.

Halting embryogenesis before it has completed the full dicot program has been an obvious evolutionary advantage to monocots, since it allows the reproductive development from flower to mature seed to take place in a much abbreviated time span.

Further comparisons of the biochemical events and their regulation between such developmental systems as cotton and barley embryogenesis and germination should provide for an even greater comprehension on a molecular level of the evolutionary events that took place as monocots emerged from dicot ancestors.

Acknowledgments

The majority of the experiments presented in this chapter were performed by Drs. James N. Ihle of the Oak Ridge National Laboratory and William C. Merrick of the National Heart Institute, National Institutes of Health, and constituted their Ph.D. research.

References

Addicott, F. T., and Lyon, J. L. (1969). *Annu. Rev. Plant Physiol.* **20,** 139.

Anderson, F. (1969). *Proc. Nat. Acad. Sci. U.S.* **62,** 565.

Brown, D. D., and Dawid, I. B. (1969). *Annu. Rev. Genet.* **3,** 127.

Chakravorty, A. K. (1968). *Biochim. Biophys. Acta* **179,** 83.

Chen, D., Sarid, S., and Katchalski, E. (1968). *Proc. Nat. Acad. Sci. U.S.* **60,** 902.

Cherry, J. H. (1968). *Symp. Soc. Exp. Biol.* **21,** 247.

Chrispeels, M. J., and Varner, J. E. (1967). *Plant Physiol.* **42,** 1008.

Davidson, E. H. (1969). "Gene Action in Early Development." Academic Press, New York.

Dure, L. S., and Merrick, W. C. (1971). *In* "Autonomy and Biogenesis of Mitochondria and Chloroplasts" (N. K. Boardman, A. W. Linnane, and R. M. Smilie, eds.), p. 413. North-Holland Publ., Amsterdam.

Dure, L. S., and Waters, L. C. (1965). *Science* **147,** 410.

Gross, P. R., and Cousineau, G. H. (1963). *Biochem. Biophys. Res. Commun.* **10,** 321.

Gross, P. R. (1968). *Annu. Rev. Biochem.* **37,** 631.

Ihle, J. N., and Dure, L. S. (1969). *Biochem. Biophys. Res. Commun.* **36,** 705.

Ihle, J. N., and Dure, L. S. (1970). *Biochem. Biophys. Res. Commun.* **38,** 995.

Ihle, J. N., and Dure, L. S. (1972a). *In* "7th International Symopsium on Plant Growth Regulators" (D. Carr, ed.), p. 216. Springer-Verlag, New York.

Ihle, J. N., and Dure, L. S. (1972b). *J. Biol. Chem.* **247,** 5034.

Ihle, J. N., and Dure, L. S. (1972c). *J. Biol. Chem.* **247,** 5041.

Ihle, J. N., and Dure, L. S. (1972d). *J. Biol. Chem.* **247,** 5048.

Ingle, J. (1968). *Plant Physiol.* **43,** 1448.

Jacobson, B. (1971). *Nature (London), New Biol.* **231,** 17.

Kirk, J. T. O., and Tilney-Bassett, A. E. (1967). "The Plastids." Freeman, San Francisco, California.

Leis, J. P., and Keller, E. B. (1970). *Proc. Nat. Acad. Sci. U.S.* **67,** 1593.

Loening, U. (1967). *Biochem. J.* **102,** 251.

Merrick, W. C., and Dure, L. S. (1971). *Proc. Nat. Acad. Sci. U.S.* **68,** 641.

Merrick, W. C., and Dure, L. S. (1972). *J. Biol. Chem.* **247,** 7988.

Ohkuma, K., Lyon, J. L., Addicott, F. T., and Smith, O. E. (1963). *Science* **142,** 1592.

Schartz, J., Meyer, R., Eisenstadt, J., and Brawerman, G. (1967). *J. Mol. Biol.* **25,** 571.

Silbert, D. F., Fink, G. R., and Ames, B. N. (1966). *J. Mol. Biol.* **22,** 335.

Skoog, F., and Armstrong, D. J. (1970). *Annu. Rev. Plant Physiol.* **21,** 359.

Steward, F., Mapes, M., and Mears, K. (1958). *Amer. J. Bot.* **45,** 705.

Stocking, C. R. (1959). *Plant Physiol.* **34,** 56.

Waters, L. C., and Dure, L. S. (1966). *J. Mol. Biol.* **19, 1.**

Watson, J. D. (1970). "Molecular Biology of the Gene," 2nd ed., p. 85. Benjamin, New York.

Weeks, D. P., and Marcus, A. (1971). *Biochim. Biophys. Acta* **232,** 671.

Wiess, J. F., and Kelmers, A. D. (1967). *Biochemistry* **6,** 2507.

3

Plant Hormones and Developmental Regulation: Role of Transcription and Translation

JOE L. KEY and LARRY N. VANDERHOEF

I. Introduction

Plant hormones are known to control or to be involved in the regulation of very diverse physiological processes. Auxin is generally considered as the hormone primarily associated with the control of cell elongation (discussed in Section III,A). Yet under many conditions auxin can also serve as the stimulus for cell division. Auxin can induce parthenocarpic fruit set and cause a normally vegetative pineapple plant to flower as other examples of the diverse activities of auxin. The cytokinins, while being considered primarily as cell division factors, affect cell enlargement, interact with auxin in the control of tissue differentiation and bud dormancy, and delay senescence. Under other conditions, cytokinins can

49

reverse the action of abscisic acid in inhibiting seed germination or in reverting abscisic acid-induced dormancy in *Lemna* (see van Overbeek, 1968). A major physiological role of gibberellin is the positive control function it exerts in the regulation of seed germination (Section III,B). Yet, gibberellins are also known to influence cell division, cell elongation, and flowering as examples of other effects it exerts on plant growth and development. Abscisic acid serves as a dormancy factor and as an "inhibitor" of many physiological processes which are enhanced by other hormones. Ethylene is involved in many kinds of growth phenomena and seems to control abscission (Section III,C), but it is also believed to induce radial enlargement of cells, mediate fruit ripening, inhibit auxin-induced growth (in association with enhancement of its synthesis by supra-optimum concentrations of auxin), and affect hook opening in seedlings. Thus, while plant biologists often view hormone action in the context of a single response (as is done in this chapter), the diverse responses elicited by a given hormone must be considered before a detailed understanding of their control of plant growth and development will be possible. In addition, the interactions of the different hormones in the control of a given physiological process add another order of complexity to an understanding of hormonal regulation. (These and other phenomena relating to hormonal regulation in plants are discussed in the selected references.)

While the internal balance of hormones may be a major controlling element in many cases where varied responses are elicited by the exogenous addition of a given hormone, it is also possible that there are several "primary" sites of action for each hormone. Another view of hormone action to account for their many and varied responses would be that the site of action is on some fundamental biochemical process such as the regulation of RNA and/or protein synthesis (transcription and translation control, respectively).

Skoog suggested in the early 1950's that the action of plant hormones related to nucleic acid metabolism. In experiments with tobacco pith callus, indoleacetic acid (IAA, the native auxin) differentially affected the RNA and DNA content of cells depending on the concentration used. Low concentrations of IAA, which induced a small amount of cell division but little cell enlargement, caused increases in DNA but not in RNA; higher concentrations of IAA increased the RNA content and cell enlargement without affecting DNA levels or cell number. In both cases the nucleic acid response to the hormone preceded the morphological response. A large number of reports on the influence of plant hormones on RNA and protein synthesis have followed (Trewavas, 1968; Key, 1969).

While there is no definitive evidence linking plant hormone action

directly to the control of either transcription or translation, there is accumulating evidence that these processes are fundamental to the hormonal control of many physiological processes. Some of the highlights
of the research on selected systems which relate hormonal regulation to
the possible control of RNA and protein biosynthesis are discussed here.

II. The Influence of Hormones on *in Vivo* RNA Synthesis

Plant hormones have been shown to enhance RNA synthesis or increase
the content of RNA in a wide variety of systems. By far the most studied
in this respect is auxin.

As indicated above auxin was shown to affect markedly the RNA content of tobacco callus in concert with alteration of the developmental
pattern of the tissue (Skoog, 1954). In general, concentrations of auxin
which enhance growth enhance RNA synthesis, and growth inhibitory
concentrations decrease RNA synthesis (see Trewavas, 1968; Key, 1969).
While these effects on RNA synthesis are usually detectable only by
radioactive precursor incorporation into RNA, auxin (usually the synthetic auxin 2,4-dichlorophenoxyacetic acid, or 2,4-D, is used) causes a
large net accumulation of RNA when applied at appropriate concentrations to intact seedlings and some excised stem tissues (e.g., Rebstock
et al., 1954; Key and Hanson, 1961; Chrispeels and Hanson, 1962; Key
et al., 1966). In intact seedlings, only those tissues where growth, albeit
abnormal, is enhanced does RNA accumulate; in the meristematic tissues
where growth is inhibited, RNA synthesis is inhibited (Key *et al.*, 1966).
The enhancement of RNA synthesis by auxin is not simply a result of
an enhanced growth rate accompanied by a general enhancement of cell
metabolism. Auxin enhances RNA synthesis within 10 to 15 min in many
tissues (e.g., Matthysse and Phillips, 1969). Also cell elongation can be
prevented in excised stem tissue by including an isotonic concentration
of some osmoticant in the incubation medium without altering the auxin
enhancement of RNA synthesis (Masuda *et al.*, 1967). Additionally,
auxin not only enhances RNA synthesis in both rapidly elongating and
fully elongated excised soybean hypocotyl, but also induces by far the
greatest enhancement of RNA synthesis in the fully elongated (i.e., mature or nonelongating) region of the hypocotyl (Key and Shannon, 1964).
The synthesis of all species of RNA (i.e., rRNA, tRNA, and the AMP-
rich components, presumably including mRNA) is enhanced by auxin
in excised tissues in relatively short-time label experiments as depicted
in Fig. 1. These data also show that the base analog, 5-fluorouracil (FU),
inhibits the accumulation of tRNA and rRNA without affecting the syn-

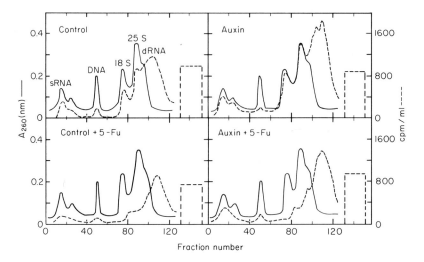

Fɪɢ. 1. Influence of auxin on RNA synthesis in control and 5-fluorouracil(FU)-treated soybean hypocotyl. RNA was extracted from elongating soybean hypocotyl, which was incubated for 4 hr in [³²P]orthophosphate following a 2-hr preincubation in 1% sucrose containing 50 μg/ml chloramphenicol and fractionated on a methylated albumin-kieselguhr column. 2,4-D was used at 10 μg/ml and 5-FU at 2.5×10^{-3} M. (From Key and Ingle, 1968.)

thesis of the AMP-rich components of RNA (Ingle *et al.*, 1965; Key *et al.*, 1972). This observation is significant since it indicates that only the AMP-rich RNA's mediate hormone-regulated processes.

As evidenced by the differential response of excised corn mesocotyl and soybean hypocotyl (see Trewavas, 1968; Key, 1969), all excised tissues do not show an enhanced accumulation of RNA or precursor incorporation into RNA in response to auxin. In the case of excised corn mesocotyl, auxin causes a large increase in RNase activity (Shannon *et al.*, 1964) and an enhanced loss of total RNA. In contrast there is no increase in RNase activity in excised soybean hypocotyl and a net accumulation of RNA occurs. This differential response of the excised tissue is not observed when intact seedlings of corn (Shannon *et al.*, 1964) and soybean (Key and Hanson, 1961) are treated with auxin. In both species there is massive accumulation of RNA. In these intact seedlings auxin treatment induces gross morphological aberrations and massive cell division (e.g., Key *et al.*, 1966). While auxin causes enhanced accumulation of all species of RNA, there are marked increases in the RNA/DNA, RNA/protein, and rRNA/tRNA ratios in auxin-treated tissue (Chrispeels and Hanson, 1962; Key *et al.*, 1966).

Recent studies by Thompson and Cleland (1971) have been directed to an analysis of the similarity of base sequences of RNA transcribed in control and auxin-treated tissues. They showed that auxin does not alter the RNA sequences being transcribed while enhancing the rate of cell elongation of excised pea tissue. This conclusion is based on the results of competition RNA–DNA hybridization experiments in which the DNA being hybridized represented primarily redundant or repeated sequence DNA (Fig. 2). This result might be expected since only the rate of cell elongation, a developmental phenomenon already initiated prior to excision of the tissue, is affected by exogenously supplied auxin, relative to that of untreated control tissue. Therefore, if RNA synthesis is an integral part of auxin-regulated cell elongation (discussed in detail in Section III,A), induction of a new family of RNA molecules may not be necessary. An enhancement of the rate of synthesis or increase in number of essential RNA transcripts might suffice to enhance the rate of cell elongation.

A different result was obtained by Thompson and Cleland when they compared RNA sequences of pea stem from control and auxin-treated intact seedlings. In this case, where auxin induces very pronounced effects on differentiation, the competition DNA–RNA hybridization experiments showed differences in the population of RNA molecules present in control and auxin-treated tissues (Fig. 3). This different response of excised and intact tissue to auxin is not understood (Thompson and Cleland, 1971),

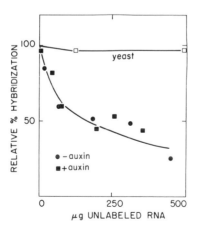

FIG. 2. Comparison by hybridization–competition of RNA species from stem sections after a 2-hr treatment with or without 5×10^{-6} M IAA. In addition to the indicated amounts of competitor RNA, each reaction contained 10 μg of reference RNA from stem sections labeled with ^{32}P for 2 hr in the presence of auxin and 20 μg of filter-bound DNA. Mixtures were incubated for 21 hr at 67°C in 2× SSC containing 0.01 M TES. (From Thompson and Cleland, 1971.)

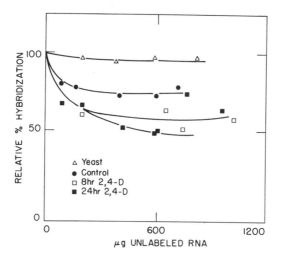

Fɪɢ. 3. Comparison of amount of new RNA species induced at 8 and 25 hr after spraying of intact pea seedlings with high levels of auxin. Reference RNA was labeled for 5 hr in ^{32}P by incubating sections excised 19 hr after spraying of seedlings with 500 μg/ml 2,4-D. Competitor RNA was obtained at 8 and 24 hr from seedlings sprayed with or without the auxin. (From Thompson and Cleland, 1971.)

but it may well relate to the deficiency of some factor or some other hormone in the excised tissue which is necessary along with auxin to elicit changes in the sequences of RNA being transcribed, and thus to further differentiation of the tissue. Possible candidates would be cytokinins and gibberellins, which are known to produce growth responses in the hypocotyl of rootless soybean seedlings when added along with auxin (e.g., Holm and Key, 1969). Cytokinins and auxin have long been known to interact in causing cell division and tissue differentiation.

In addition to affecting RNA synthesis, auxin also induces changes in protein synthesis and in patterns of protein synthesis. There are many noted changes in enzymatic activities (e.g., Venis, 1964; Fan and Maclachlan, 1967; Stafford and Galston, 1970), in protein bands as observed on polyacrylamide gels (e.g., Morris, 1966) and in total protein content following auxin treatment. Even though many of these changes are dependent on RNA synthesis, the relationship between auxin-enhanced RNA synthesis and changing protein synthesis has not been studied in detail. Since many of the changes in protein have been observed in seedlings treated with auxin, the changes in the population of RNA molecules noted by Thompson and Cleland for the pea undoubtedly are associated directly with the modified protein patterns.

While auxin has been shown by many laboratories to increase the synthesis of AMP-rich RNA's, presumably including mRNA (see Trewavas, 1968; Key, 1969), work with polyribosomes also indicates an effect of auxin on the synthesis and/or utilization of mRNA. Excising pea internode sections results in a loss in total ribosomes, with the loss occurring in the polyribosome fraction (Trewavas, 1968). With auxin present, there is a small increase in ribosomes and a large increase in the proportion of ribosomes present as polyribosomes. Similarly in excised elongating soybean hypocotyl, there is a net loss of ribosomes and a marked decrease in the relative level of polyribosomes (Anderson, 1972). In this tissue, however, auxin has little or no effect on the proportion of ribosomes present as polyribosomes. In contrast to the elongating tissue, in the fully elongated (mature) soybean hypocotyl auxin induces an RNA synthesis-dependent transformation from 35 or 40% polyribosomes to about 75% polyribosomes. Thus, either the synthesis and/or utilization of mRNA must be altered by auxin.

Although not studied as much as auxin, other hormones are known to affect RNA synthesis. Gibberellin (GA) also causes marked changes in RNA metabolism. As examples, in the barley aleurone system detailed in Section III,B, GA enhances precursor incorporation into RNA early after treatment (Varner *et al.*, 1965; Chandra and Varner, 1965) and prior to the onset of GA-induced RNA degradation. The GA-induced loss of RNA after about 24 hr apparently relates to the GA-regulated increase in RNase activity (Chrispeels and Varner, 1967a). There was suggestive evidence that GA may have affected the kinds as well as the amount of RNA being synthesized. In dwarf pea plants which show marked growth responses to GA, there is an associated increase in RNA prior to the enhancement of DNA synthesis and cell division (e.g., Broughton, 1968). In association with the retardation of senescence of leaf tissues, GA (Beevers, 1966; Fletcher and Osborne, 1966) and cytokinin (e.g., Osborne, 1965) have been shown to enhance RNA synthesis (discussed in Section III,D).

III. RNA and Protein Synthesis and the Expression of Hormone Action

In this discussion, only one of the major physiological responses to each class of plant hormones will be covered: the auxin control of cell elongation, the gibberellin (and abscisic acid) regulation of hydrolase synthesis in germinating seed, the ethylene control of abscission, and the cytokinin (and GA) retardation of senescence.

A. AUXIN CONTROL OF CELL ELONGATION

Although auxin is involved in the regulation of many physiological processes in plants, the control of cell elongation represents the most studied auxin response. The evidence that continued RNA and protein synthesis are essential for continued cell elongation comes primarily from the use of specific inhibitors of the biosynthesis of these macromolecules (actinomycin D and cycloheximide which directly inhibit RNA synthesis and protein synthesis, respectively, are the most widely used inhibitors). Early work by Masuda (1959), Nooden and Thimann (1964), and Key (1964) showed that RNA and protein synthesis were essential for continued cell elongation in excised tissues. Similar observations have been made for the inhibition of cell elongation in intact seedlings where both actinomycin and cycloheximide dramatically inhibit cell elongation in concert with the inhibition of RNA and protein synthesis (Lin and Key, 1968; Holm and Key, 1969). The inhibition by actinomycin of cell elongation of tissue growing at some preestablished steady state rate (independent of the growth rate, i.e., ranging from the low endogenous growth rate in the absence of added auxin up to the maximum rate in the presence of optimum auxin) occurs with a lag of 1 to 2 hr (Fig. 4A). This lag does not relate to a slow penetration of actinomycin and thus a delayed inhibition of RNA synthesis (Fig. 4B). Clearly, the inhibition of cell elongation by actinomycin is preceded by the inhibition of RNA synthesis. After the initial lag in the inhibition of cell elongation a new

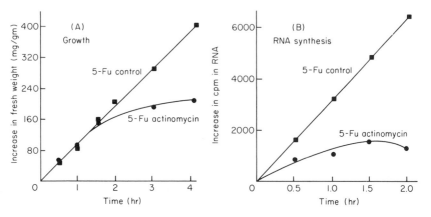

FIG. 4. Timing of inhibition by actinomycin D of growth and RNA synthesis of 5-fluorouracil-treated soybean hypocotyl. One-gram samples of soybean hypocotyl (elongating region) were pretreated for 2 hr in medium containing 10 μg/ml 2,4-D, 325 μg/ml 5-fluorouracil, and 0.5 μCi [8-,^{14}C]ADP. At zero time actinomycin D (10 μg/ml) was added to one series of flasks.

steady state growth rate is established commensurate with the level of RNA synthesis over a wide range of actinomycin concentrations (e.g., Key *et al.*, 1967). The failure of FU to inhibit auxin-induced cell elongation (Key and Ingle, 1964), while dramatically inhibiting rRNA and tRNA synthesis (see Fig. 1), shows that only the synthesis of AMP-rich RNA is required to support endogenous and auxin-induced cell elongation. There is complete parallelism between the inhibition of auxin-induced cell elongation and the synthesis of AMP-rich RNA by a wide range of actinomycin concentrations (Fig. 5), even though the preestablished control or endogenous growth continues at a measurable rate for a few hours. Cycloheximide likewise leads to a parallel inhibition of auxin-induced cell elongation and protein synthesis over a wide concentration range (Fig. 6). As would be expected, actinomycin inhibition of RNA synthesis leads to an inhibition of protein synthesis (Key *et al.*, 1967). Similar to the effects of RNA synthesis, actinomycin inhibits a given increment of protein synthesis at the same time a corresponding increment of auxin-induced growth is inhibited. Various kinetic measurements of the growth rate in connection with actinomycin and cycloheximide treatment indicate that the concentrations of "growth-essential" RNA and protein are increased by auxin (Key *et al.*, 1967; Key and Ingle, 1968; Cleland, 1971b). Finally the results to date generally indicate that auxin cannot enhance the rate of cell elongation in tissue where RNA and protein synthesis are previously inhibited, thus indicating that

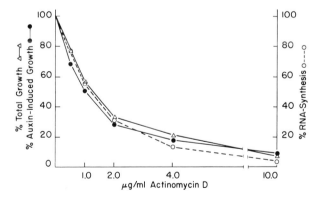

Fig. 5. Parallel inhibition by actinomycin D of auxin-induced growth and AMP-rich RNA synthesis. One-gram samples of elongating soybean hypocotyl were pre-incubated for 4 hr in basic medium, 2.5×10^{-3} M 5-fluorouracil and the various concentrations of actinomycin D. After 4 hr auxin (10 μg/ml 2,4-dichlorophenoxy-acetic acid) and 0.5 μCi [8-^{14}C]ADP were added with auxin-induced growth and RNA synthesis being measured over the next 4-hr interval. (From Key *et al.*, 1967.)

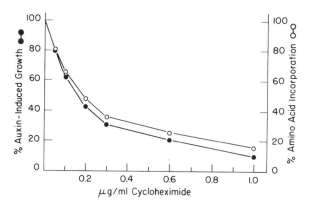

Fig. 6. Parallel inhibition of auxin-induced growth and protein synthesis by cycloheximide. One-gram samples of elongating soybean hypocotyl were preincubated for 4 hr in the indicated concentrations of cycloheximide prior to the addition of auxin (10 μg/ml 2,4-dichlorophenoxyacetic acid) and 0.5 μCi [¹⁴C]leucine. Auxin-induced growth and protein synthesis were measured over the next 4-hr interval. (From Key *et al.*, 1967.)

auxin does not amplify preexisting RNA and protein in causing an enhancement of the rate of cell elongation. There are, however, some conflicting reports on this latter point, particularly in studies using pea epicotyl (Penny and Galston, 1966; de Hertogh *et al.*, 1965; Barkley and Evans, 1970). The different laboratories report very different initial growth rates in response to auxin following actinomycin pretreatment. However, no monitoring of RNA synthesis inhibition by actinomycin was done in these experiments.

The results presented above and those reviewed by Trewavas (1968 and Key (1969) indicate a very close connection between the enhancement of RNA synthesis by auxin, and in turn an increased synthesis of "growth-essential" protein, and the auxin control of cell elongation. There are, however, reports of very fast growth responses to auxin which probably cannot be mediated by RNA and protein synthesis. A typical growth response curve by *Avena* coleoptile to auxin is shown in Fig. 7. There is a definite lag of 10 to 12 min followed by a very fast attainment of a new steady state growth rate. Evans and Ray (1969) showed that pretreatment of tissue with actinomycin and cycloheximide did not lengthen this lag before attainment of the new steady state growth rate, although this rate was considerably lower than in untreated controls. Based on this observation and a kinetic analysis of their data, Evans and Ray concluded "that auxin probably does not act on the elongation of these tissues (*Avena* and corn coleoptiles) by promoting the synthesis of informational RNA or enzymatic protein."

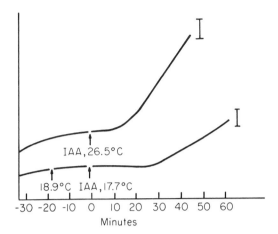

Fig. 7. Timing of the auxin response at 17.7°C as compared with 26.5°C. Upper curve: Medium changed from water at 26.5°C to 3 μg/ml IAA at 26.5°C at arrow. In lower curve, medium changed from water at 26.5°C to water at 18.9°C (first arrow) and then to IAA at 17.7°C (second arrow). In this experiment each curve was obtained using 10 coleoptile segments. (From Evans and Ray, 1969.)

More recently, Rayle *et al.* (1970a) showed that lag times for auxin-enhanced cell elongation as short as 2 or 3 min could be observed by careful manipulation of the IAA concentration or by use of the appropriate concentration of the methyl ester of IAA. They also reported a transient inhibition of the rate of cell elongation which occurred in 1 to 3 min. These fast growth responses following auxin treatment are not unique to auxin. A shift from a near neutral pH to a pH of about 3.5 in the incubation medium leads to a rapid enhancement of the rate of cell elongation independent of auxin (e.g., Rayle *et al.*, 1970b; Hager *et al.*, 1971). Likewise a high concentration of CO_2, independent of a pH change, causes a rapid enhancement of cell elongation (Rayle and Cleland, 1970; Evans *et al.*, 1971). A number of lipophilic agents such as long-chain alcohols are known to cause an enhancement of the rate of cell elongation. The rapid response is possibly indicative of an effect of these compounds on the cell membrane, an idea entertained by Veldstra relative to the action of auxin as early as 1953.

While these fast growth responses to auxin, pH drop, high CO_2 concentration, and various lipophilic agents are of interest in themselves, there is as yet no definitive evidence as to how they relate to the normal processes of auxin-regulated cell elongation. While these fast responses probably relate to some physical modification of the cell membrane, it is entirely conceivable that they are not necessarily directly related to the normal auxin control of cell elongation. This would suggest, possibly, that it is coincidental that auxin exerts this fast response based on its lipophilic properties while normally regulating cell elongation through other processes. That the fast growth response may indeed be different from the sustained steady state growth rate which occurs in response to auxin comes from experiments with tissues other than *Avena* coleoptile, e.g., pea (Rayle *et al.*, 1970a), in which the rate of growth in the initial fast response is considerably greater than the steady state rate which is attained in about 15 to 20 min. A sufficient number of plant tissues have not been studied to know whether this response in pea is a general phenomenon. See Ray (1969) and Cleland (1971a) for a detailed discussion of some of the phenomena mentioned above.

It is apparent that the mechanism(s) by which auxin regulates cell elongation is not understood. The fast responses, i.e., 1 to 3 min, are undoubtedly not mediated via an effect of auxin on either transcription or translation, nor is the fast response auxin-specific. On the other hand, the sustained steady state growth in response to auxin does require the participation of RNA and protein synthesis, and the level of "growth-essential" RNA and protein is increased by auxin. Thus, at this stage in our understanding of the mechanism(s) through which auxin regulates cell elongation, the rapid response to auxin (and other agents) and the influence of auxin on RNA synthesis should not be viewed as separate phenomena relative to auxin action. Therefore, the idea that auxin action involves an either/or situation should not be emphasized; instead, their integration into the total framework of the physiology of auxin regulation should be sought.

B. GIBBERELLIN REGULATION OF HYDROLASE SYNTHESIS IN GERMINATING BARLEY SEED

The hormonal control of enzyme synthesis, primarily enzymes involved in food reserve mobilization, has been studied extensively in germinating barley seeds. Gibberellin (GA) is directly involved in the positive control ("induction") of synthesis of many hydrolases in the barley system (e.g., Varner and Chandra, 1964). Abscisic acid (ABA) appears to function as a negative control ("repression") in this system, and in fact will inhibit the GA-enhanced synthesis of the hydrolase enzymes (e.g., Chrispeels and Varner, 1966). The complexity of the hormonal control in this system is further evidenced by the fact that cytokinins can reverse the ABA inhibition of both seed germination and α-amylase (the hydrolase most studied in the barley system) synthesis in barley (Khan, 1971). Khan assigns a "permissive" role for cytokinins and a "preventive" role to ABA with GA assuming the primary role in the control of seed germination. The synthesis of the hydrolases in barley requires the participation of RNA synthesis presumably including mRNA during the germination phase (e.g., Varner and Chandra, 1964). This contrasts the "germination" enzymes in cotton seed where the associated mRNA is synthesized during embryogenesis but only translated during germination (see Chapter 2 by Dure in this volume for a detailed discussion of this system). Only pertinent aspects of the control of hydrolase synthesis in the germinating barley seed will be discussed here.

Some 80 years ago Haberlandt (1890) discovered that the aleurone cells of rye seeds produced substances which caused liquefaction of the starchy endosperm and dissolution of the starch grains. A viable embryo was required for liquefaction to proceed. Some years later Brown and Escombe (1898) made a similar observation in barley. Paleg (1960) and Yomo (1961) independently discovered that GA treatment of barley seeds caused an increase in amylolytic enzymes and a release of reducing sugars. It was then suggested that GA caused an activation of "inactive" amylase present in the seed. It is now known, however, that GA causes a *de novo* synthesis of α-amylase and other hydrolases in the aleurone cells of the seed (Filner and Varner, 1967; Jacobson and Varner, 1967). GA which arises in the embryo portion of the seed during imbibition is subsequently translocated through vascular connections to the aleurone cells (MacLeod, 1966) where the enzymes are synthesized. That the aleurone layer is the site of α-amylase production is shown in Fig. 8.

The demonstration of *de novo* synthesis of the hydrolase enzymes (e.g., α-amylase and protease) in the barley aleurone cells comes from a series of elegant studies in Varner's laboratory. Varner (1964) showed that

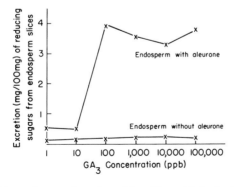

FIG. 8. Induction of α-amylase by gibberellic acid in barley endosperm tissue. The aleurone cells are shown to secrete the enzymes that hydrolyze starch to reducing sugar. (From van Overbeek, 1966.)

there was marked incorporation of ^{14}C-labeled amino acid into a protein band (obtained by DEAE-cellulose column fractionation) containing the α-amylase activity of extracts only from tissue which had been incubated in the presence of GA. Inhibitors of protein synthesis also prevented the appearance of α-amylase activity in response to GA. Fingerprints of tryptic digests of purified α-amylase produced in the presence of GA showed that the [14C]threonine was incorporated throughout the molecule (Varner and Chandra, 1964). Knowing that at least some of the increase in α-amylase activity which occurred in response to GA was the result of *de novo* synthesis, Filner and Varner (1967) then conducted density labeling experiments in conjunction with isopycnic equilibrium centrifugation to establish whether all of the increase in α-amylase activity was the result of *de novo* synthesis of GA. The density labeling of the amylase depended on the *in vivo* generation of ^{18}O-labeled amino acids derived from reserve protein of the seed and $H_2^{18}O$ of the medium around the aleurone layers as depicted in the following equation:

$$\text{Reserve protein} + H_2^{18}O \xrightarrow{\text{proteolysis}} RCHNH_2C^{16}O^{18}OH + RCHNH_2C^{18}O^{16}OH$$

Newly synthesized proteins containing ^{18}O-labeled amino acids would thus have a higher density than those synthesized from normal ^{16}O-amino acids. The gradient distribution of α-amylase activity extracted from aleurones incubated in $H_2^{18}O$ and GA relative to those incubated in $H_2^{16}O$ (Fig. 9) demonstrated that essentially all of the increase in α-amylase was the result of *de novo* synthesis from amino acids derived from stored

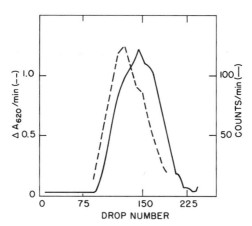

Fig. 9. Equilibrium distributions of radioactivity in α-amylase (———) and α-amylase enzyme activity (- - - -) after centrifugation of a mixture of 2 μg of purified [16O³H]α-amylase and about 25 μg of crude α-amylase induced by GA in $H_2^{18}O$. (From Filner and Varner, 1967.)

protein in response to GA. Again using this technique, Jacobson and Varner (1967) showed that protease is also synthesized *de novo* in response to GA.

There is a lag of about 8 hr following GA treatment before there is an increase in amylase activity in the barley half-seed (i.e., less the embryo-containing portion of the seed) or isolated aleurone layers (Varner and Chandra, 1964). This lag can be shortened by extending the period of imbibition prior to the addition of GA. The removal of GA in "midcourse" results in a rapidly diminishing rate of increase in amylase activity (Chrispeels and Varner, 1967b). Thus GA does not act as a "trigger" for enzyme synthesis but is required throughout the period of enzyme production. Reintroduction of GA into the barley aleurones, in which the rate of amylase production was greatly diminished (e.g., by "midcourse" removal of GA), results in an increase in amylase production without a lag at a rate equivalent to that under continuous GA treatment.

Actinomycin D prevents amylase synthesis when added before or at the same time as the hormone, indicating a requirement for RNA synthesis in the GA-mediated synthesis of amylase (Varner *et al.*, 1965; Chrispeels and Varner, 1967b). If added 8 hr or more after GA, actinomycin is essentially without effect on enzyme synthesis even though RNA synthesis is decreased by about 65%. The synthesis of amylase which depends upon readdition of GA following "mid-course" removal is likewise not very sensitive to actinomycin D. The base analogs, 6-methylpurine and 8-azaguanine, effectively inhibit amylase synthesis after a lag of 2 to 3 hr, independent of the timing of their addition (Fig. 10). As is the case for cell elongation in response to auxin, the base analog, 5-fluorouracil, is without effect on GA-induced hydrolase synthesis.

As pointed out previously, another most interesting aspect of the hormonal regulation of amylase synthesis in the barley aleurone relates to the effects of ABA (Chrispeels and Varner, 1966, 1967b). ABA inhibits the GA-induced synthesis of amylase without significantly affecting amino acid incorporation into total protein and only slightly depressing RNA synthesis (Chrispeels and Varner, 1966). That the ABA inhibition of GA-induced enzyme synthesis is highly specific is further indicated by the fact that the substrate induction of nitrate reductase in the barley aleurone is only slightly inhibited by ABA while amylase synthesis is inhibited by about 90% (Ferrari and Varner, 1969). Although inhibition of amylase synthesis by ABA can be partially overcome by increasing the concentration of GA, the inhibition does not appear to be strictly competitive.

The midcourse addition of ABA results in the inhibition of amylase synthesis with essentially the same kinetics as inhibition by base analogs

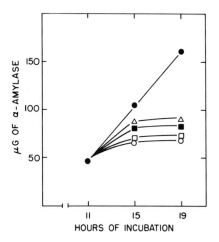

FIG. 10. Midcourse inhibition of α-amylase synthesis by abscisic acid, 6-methyl-purine, and 8-azaguanine. Aleurone layers were incubated in 0.05 μM GA for 11 hr. At this time the medium was removed, the aleurone layers were rinsed, then they were further incubated with 0.05 μM GA (●——●), or with GA and 10 μM abscisic acid (○——○), 5 mM 6-methylpurine (□——□), 0.5 mM 6-methylpurine (■——■), or 5 mM 8-azaguanine (△——△). (From Chrispeels and Varner, 1967b.)

(Fig. 10). This is suggestive of the possibility that ABA may inhibit GA-mediated hydrolase synthesis by inhibiting the synthesis of a specific RNA fraction required for their synthesis (Chrispeels and Varner, 1967b). There is, however, no evidence for a direct effect of either GA or ABA on these specific RNA's. In summarizing their results on the GA- and ABA-regulated synthesis of hydrolases in the barley aleurone, Chrispeels and Varner (1967b) stated, "It is not yet possible to decide whether these hormones work at the level of transcription or at the level of translation. The data . . . are consistent with the hypothesis that GA exerts its control at the level of the gene, to bring about the synthesis of an RNA fraction specific for the proteins being synthesized. (Abscisic acid would be assumed to repress this synthesis.) The data are equally consistent with a control mechanism at the level of translation with a requirement for continued RNA synthesis."

While GA induces RNA synthesis-dependent *de novo* synthesis of hydrolases in barley aleurones, the mechanism(s) through which GA exerts this action remains obscure. GA also affects secretion of hydrolases in addition to affecting their synthesis (Chrispeels and Varner, 1967a). There are also many structural and biochemical changes which are now known to occur during the lag period prior to the appearance of α-amylase, protease, and other hydrolase activities. One of these must be the proteolysis of stored protein since the density labeling of GA-induced enzymes depended on this hydrolysis (Filner and Varner, 1967). Jones (1969a) has interpreted the inhibition of GA-induced α-amylase synthesis by mannitol and polyethylene glycol to be the result of a reduced proteolysis of stored protein of the aleurone grain thus making available less substrate for new protein synthesis. Protein synthesis per se is not inhibited by the osmoticants. There is also a GA-mediated early release or secretion of a β-1,3-glucanase in the aleurone layers (Jones, 1971).

At the ultrastructural level, there is extensive development of rough endoplasmic reticulum (ER) as well as changes in the rough ER of isolated aleruone layers during the lag and synthesis phases of α-amylase production (Eb and Nieuwdorp, 1967; Jones, 1969b,c). Evins and Varner (1971) have recently shown that GA increases the rate of synthesis of ER by four- to eightfold based on [^{14}C]choline incorporation into an ER-rich membrane fraction within 4 hr. This coincides with a GA-induced increase in the number and proportion of ribosomes which can be isolated as polysomes, starting about 3 hr after hormone treatment (cited in Evins and Varner, 1971). These effects precede and are likely required for the GA-induced synthesis and release of hydrolytic enzymes. ABA prevents the GA-enhanced synthesis of ER and the increase in ER-bound polyribosomes. Actinomycin and cycloheximide also prevent the increase in the rate of ER synthesis.

Even with the many significant findings discussed above, the exact mechanisms through which the hormones exert their control over seed germination remain to be elucidated. There seems little doubt, however, that both transcription and translation (also see Dure, Chapter 2) events are hormonally regulated in the control of seed dormancy and germination. A model for hormone interactions in these processes was presented by Khan (1971).

C. ETHYLENE-REGULATED ABSCISSION

Ethylene is known to enhance the rate of abscission in appropriately aged tissue explants. Since there are no known ways of specifically blocking endogenous ethylene production, it has not been possible to ascertain if this endogenously produced ethylene normally regulates abscission. The endogenous level of ethylene production is certainly sufficient eventually to cause abscission, and a reasonable interpretation seems to be that ethylene is a natural regulator of abscission.

Abeles and Holm (1966) showed that ethylene enhances RNA and protein synthesis in explants which are aged to the state where they abscise in response to applied ethylene (Fig. 11). The enhancement of RNA syn-

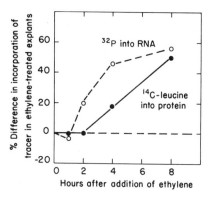

Fig. 11. Time course of enhancement of [^{32}P]orthophosphate and [^{14}C]L-leucine incorporation into RNA and protein of ethylene-treated vs. control bean explants. Agar blocks of ^{32}P and [^{14}C]L-leucine were placed on top of senescent explants, and the explants were incubated in either air or an atmosphere containing 4 ppm ethylene. The RNA and protein was extracted at 2, 4, 6, and 8 hr. (From Abeles and Holm, 1966.)

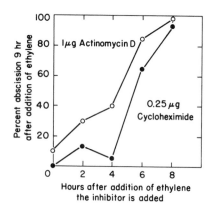

Fig. 12. Effect of actinomycin D(○) and cycloheximide (●) on abscission of senescent bean explants. Actinomycin D and cycloheximide were injected into the ethylene-treated kidney bean explants in 1 μl of solution. Injections were made at 0, 2, 4, and 6 hr. Abscission was measured 9 hr after the start of the experiment. (From Abeles and Holm, 1966.)

thesis precedes the increase in protein sythesis, with both preceding noticeable evidence of an effect on abscission proper. The ethylene-enhanced synthesis of RNA and protein occurs primarily in the abscission zone of explants, but not in the regions of the explant away from the abscission zone.

The ethylene enhancement of the rate of abscission is effectively blocked by treatment with either actinomycin or cycloheximide (Fig. 12) when added prior or up to about 4 hr after the addition of ethylene. These results show a requirement for RNA and protein synthesis in the abscission process. This, taken with the enhancement of both RNA and protein synthesis, commmensurate with the enhanced rate of abscission, is consistent with the primary role of ethylene being to bring about specific RNA synthesis necessary for the synthesis of enzymes essential for cell wall digestion in the separation layer (Abeles and Holm, 1966). A possible enzyme candidate for this role is cellulase. Abeles (1969) showed that ethylene causes an increase, after about a 3-hr lag, in the level of cellulase activity in the abscission zone. Actinomycin and cycloheximide blocked the ethylene-induced increase in cellulase activity, suggestive of *de novo* synthesis of the enzyme in response to ethylene. By using the density labeling technique (Section III,B), Lewis and Varner (1970) demonstrated *de novo* synthesis of cellulase in bean explants. Abeles (1969) concluded that one of the roles of ethylene in abscission is to regulate the synthesis of cellulase required for cell separation, which is presumably regulated by ethylene-mediated RNA synthesis.

D. Retardation of Senescence by Hormones

The work of Richmond and Lang (1957) showed that kinetin (a synthetic cytokinin, see Section IV) retarded protein loss in senescing leaves while delaying senescence (usually monitored by chlorophyll loss). This observation led to considerable work on the effects of cytokinins and other hormones on nucleic acid and protein metabolism in senescing tissues (e.g., Beevers, 1966; Fletcher and Osborne, 1966; Osborne, 1965; Srivastava and Ware, 1965; Srivastava, 1968; Wollgiehn, 1967; Sacher, 1967). Osborne's results showed that kinetin delayed or slowed the loss of both DNA and RNA which normally accompanies senescence. The kinetin-enhanced incorporation of ^{14}C-labeled precursors into RNA and protein caused Osborne (1965) to suggest that the effect of kinetin in retarding senescence was through its action in sustaining nucleic acid and protein synthesis. The enhanced incorporation of precursors into RNA and the associated maintenance of a higher RNA content in response to both GA (Beevers, 1966; Fletcher and Osborne, 1966) and auxin (Sacher, 1967) while delaying senescence in a variety of plant tissues generally supports Osborne's view of the mechanism by which hormones delay senescence. However, from their studies on the influence of auxin and kinetin on RNA synthesis and senescence in detached leaves, von Abrams and Pratt (1968) concluded that hormones do not regulate senescence by an effect on RNA synthesis.

The hormone-mediated maintenance of the RNA and protein contents in tissues where senescence is delayed may be mediated via effects on nuclease and protease activities instead of (or in addition to) affecting the synthesis of these macromolecules. Srivastava and Ware (1965) showed that kinetin-treated barley leaves contained less RNase and DNase activities than control tissue. These nuclease activities associated with chromatin isolated from excised barley leaves increased several times during 4 days of normal senescence (Srivastava, 1968). Kinetin prevented this increase in chromatin-associated nuclease activity. Kinetin also suppresses the increase in protease activity which normally accompanies senescence of leaves (e.g., Balz, 1966). These effects of kinetin on nuclease and protease activities may have a twofold influence on the observed enhancements of radioactive precursor incorporation into RNA and protein while delaying senescence. First, less newly synthesized RNA and protein would be degraded in hormone-treated tissue. The results of Kuraishi (1968) and of Tavares and Kende (1970) show that ^{14}C-labeled protein degradation is in fact lower in cytokinin-treated than in control tissue. Second, the precursor pools would be expected to be larger in control than in hormone-treated tissue because of a retarded macromolecule breakdown to precursors in the treated tissue, thus causing less dilution of the absorbed radioactive precursor in the hormone-treated tissue. Again this possibility is supported by measurements of total α-amino nitrogen in senescing leaf discs (Anderson and Rowan, 1966; Tavares and Kende, 1970). Tavares and Kende (1970) thus suggested that cytokinins retard senescence primarily through inhibiting protein degradation rather than through an enhancement of synthesis. The possibility that the hormone-mediated effects on hydrolytic enzyme activities and on radioactive precursor incorporation may be secondary to the retardation of senescence by the hormones should also be considered.

IV. The Occurrence of Cytokinins in Transfer RNA

Since the discovery that kinetin (see Fig. 13 for structure of some synthetic and naturally occurring cytokinins) was the active cell division factor of old DNA preparations, a large number of 6-substituted aminopurines has been shown to possess cytokinin activity (i.e., to induce cytokinesis or to cause some other response such as senescence retardation). Among these compounds are isopentenyladenine (iP) and its riboside (iPA). Zeatin, dihydrozeatin, and their ribosides and ribonucleotides are naturally occurring cytokinins and similar in structure to the iP group [see Helgeson (1968) and Skoog and Armstrong (1970) for more detail on active cytokinins].

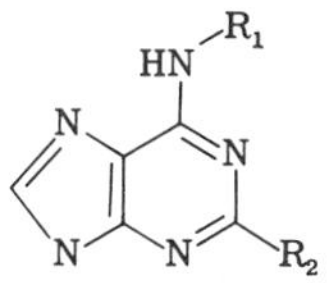

Kinetin: R_1 = furfuryl-, R_2 = H
Benzyladenine: R_1 = benzyl-, R_2 = H
Zeatin: R_1 = 3-hydroxymethyl-2-butenyl-,
   R_2 = H
Isopentenyladenine: R_1 = isopentenyl-,
   R_2 = H
2-Thiomethyl-cis-zeatin: R_1 = 3-hydroxy-
   methyl-2-butenyl-, R_2 = —SCH_3
2-Thiomethylisopentenyladenine: R_1 = iso-
   pentenyl-, R_2 = —SCH_3

Fig. 13. Some natural and synthetic cytokinins. All the cytokinins, with the exception of kinetin and benzyladenine, are naturally occurring.

While there were some early indications that cytokinins might function to regulate protein synthesis (see Section III,D), a new excitement about the possible functioning of cytokinins in the control of protein synthesis came with the discovery that bases with cytokinin activity occur naturally in tRNA. Zachau and co-workers (1966) identified iP as a constituent of two serine-accepting tRNA's of yeast. The same base was found by Madison and Kung (1967) in a tyrosine tRNA. From the sequence analysis of these tRNA's it was shown that this cytokinin-active base occurred adjacent to the 3′ end of the anticodon (Fig. 14). It now seems likely that some tRNA's of all organisms contain bases which possess cytokinin activity. Zeatin riboside is the predominant type of cytokinin in tRNA preparations from higher plants; in addition iPA and the

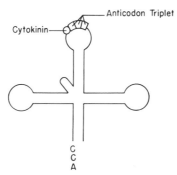

Fig. 14. Location of the cytokinin in the primary sequence of transfer RNA. The cytokinin base is located on the 3′ side of the anticodon. It is only found in tRNA's which respond to UXX codons. All such tRNA's, however, do not contain cytokinin.

2-methylthio derivatives of both iPA and zeatin riboside have also been identified (e.g., Hecht *et al.*, 1969; Burrows *et al.*, 1970). The modified base possessing cytokinin activity apparently occurs only in those tRNA's which respond to codons beginning with U (e.g., serine, leucine, tyrosine, cysteine, and tryptophan) and only adjacent to the anticodon; however, not all tRNA's which respond to code words beginning with U contain a cytokinin-active base. (See Skoog and Armstrong, 1970, for details of the work on occurrence of cytokinins in tRNA.)

The location of the cytokinin base adjacent to the anticodon was suggestive of a possible role of the cytokinin in the functioning of the tRNA. That the cytokinin base is involved in the functioning of a seryl-tRNA was reported by Fittler and Hall (1966). They showed that the binding of a yeast seryl-tRNA to ribosomes was severely impaired by modification of the iPA within this tRNA by treatment with aqueous iodine. The ability of this tRNA to accept serine, however, was not altered. The functioning of a phenylalanyl-tRNA which contained no cytokinin base was not affected by identical treatment. Since the iodine treatment only modified the iPA rather than removing the isopentenyl group, definitive conclusions on the functioning of the base were not possible. Gefter and Russell (1969) presented good evidence that the cytokinin bases do serve an important role in the functioning of tRNA. By using *E. coli* infected with a defective transducing phage carrying the tyrosine tRNA gene, selective synthesis of tyrosine tRNA was achieved (for convenience these experiments utilized a mutant tyrosine tRNA gene, i.e., an amber suppressor). Three forms of suppressor tyrosine tRNA were isolated and found to differ only in the modification of the adenine residue adjacent to the anticodon (i.e., unmodified adenine, iPA, or methylthio-iPA). These three tRNA's were acylated with tyrosine at the same rate. However, they differed significantly in their binding to ribosomes and in *in vitro* tests of suppression. The tRNA containing unmodified adenine adjacent to the anticodon was ineffective in these tests relative to the molecules containing iPA or the methylthio-iPA. These results along with those from Hall's group suggest that the cytokinin bases in tRNA have important roles in the regulation of protein synthesis. There are, however, two lines of evidence which raise some questions about the role of the iPA or methythio-iPA adjacent to the anticodon in the functioning of the aminoacyl-tRNA in protein synthesis. First, Litwack and Peterkofsky (1971) have shown that mevalonic acid, an essential growth factor for *Lactobacillus acidophilus* 4963, is the precursor of the isopentenyl group of iPA in the tRNA. By controlling the mevalonic acid concentration, the requirement for growth was met while the tRNA contained only about one-half the maximum iPA content. The iPA-deficient tRNA was indistinguishable from iPA-saturated tRNA in aminoacylation as well as in *in vitro* protein synthesis. Similarly, *Mycoplasma sp* (Kid) tRNA which contains few minor bases and apparently no iPA can function in an *E. coli* protein-synthesizing system with natural or synthetic message. Thus, there is some question about the functional significance of iPA or a similar cytokinin in tRNA relative to protein synthesis. A role for the cytokinin base in some regulatory function of aminoacyl-tRNA possibly should be considered. It is also of possible interest to note that benzyladenine (an active cytokinin, Fig. 13) treatment of soybean seedlings caused changes in the relative abundance of certain leucine tRNA's (Anderson and Cherry, 1969). Whether these changes have a functional significance is not known at this time.

While the cytokinin bases of tRNA may be of importance to the proper functioning of the tRNA in protein synthesis, it is not at all established that this phenomenon relates to the mechanism of action of cytokinins as plant hormones. Some of the relevant considerations are discussed below. Base modification in tRNA normally occurs after the primary structure is transcribed. In the case of 2iPA, mevalonic acid serves as a precursor *in vivo*, and there is some evidence for *in vitro* incorporation into tRNA (see Skoog and Armstrong, 1970). The experiments to date do not rule out the possibility that free cytokinins may serve as precursors to, or otherwise be involved in the formation of, the cytokinin bases in tRNA. Incorporation of benzyladenine (or 6-benzylamino-purine) into RNA of callus was demonstrated by Fox (1966) and Fox and Chen (1967). Methylene-labeled benzyladenine was incorporated primarily into tRNA with a significant amount of the label being recovered as benzyladenine from hydrolysates of the tRNA. Complete base substitution in the tRNA or side-chain transfer to adenine residues in tRNA, however, could account for the observed incorporation. Kende and Tavares (1968) also demonstrated incorporation of ^{14}C from methylene-labeled benzyladenine into soybean callus tRNA, but were unable to detect incorporation of ^{14}C from 6-benzylamino-9-methylpurine labeled in the same position. While this compound promoted callus growth, it could not be involved in riboside formation, and thus incorporation into tRNA, without prior cleavage of the 9-methyl group (lability of this group has been reported and may account for the biological activity of this molecule). In the same study Kende and Tavares (1968) were unsuccessful in attempts to isolate cytokinin-requiring mutants of *E. coli*, and thus concluded that it was unlikely that cytokinins exert their regulatory role as plant hormones through incorporation into tRNA.

Burrows *et al.* (1971) have identified the cytokinin bases that occur in tRNA of the cytokinin-requiring tobacco callus when grown on benzyladenine. The tRNA was extracted from the callus, purified, and hydrolyzed to constituent nucleosides. The hydrolysates were fractionated and cytokinin activity was assayed. In conjunction with these assays, mass spectral analysis, chromatography, and ultraviolet spectra demonstrated the presence of four cytokinins in the tRNA. Benzyladenosine, 2iPA, zeatin riboside, and the methylthio derivative of the zeatin riboside were identified. Thus, while benzyladenine was incorporated into the tRNA, the major cytokinin bases present in the callus tRNA were those normally found in tRNA's of other plants and not just the exogenous cytokinin upon which callus growth depended. It was suggested that the biosynthesis of the "natural" cytokinin ribonucleosides is a part of the increased anabolic activity induced by the exogenous supply of cytokinin.

Brandes and Kende (1968) also concluded that cytokinin action did not involve covalent bond formation into some plant constituent, such as tRNA, based on a series of experiments with moss protonema. Cytokinins induce bud formation in the protonema, eventually leading to branch formation. When grown on radioactive benzyladenine, the hormone was concentrated in the target cells. The removal of external cytokinin and washing of the tissue resulted in loss of most of the radioactivity from these cells, the bud formation ceased, and filamentous growth continued. While possibly suggestive of a noncovalent interaction of cytokinins in this response, the possibility that cytokinin-containing macromolecules such as tRNA might be broken down (turnover) in 18 hr must not be overlooked; a "regulatory" macromolecule might well be expected to be labile and possess a relatively short half-life.

While an importance of cytokinins in tRNA functioning is suggested, considerably more work is needed to show whether cytokinin action as a plant hormone relates to its presence in tRNA. In concluding their review, Skoog and Armstrong (1970) wrote, "as a hopeful approach to this problem (i.e., cytokinin action as a plant hormone), it is now possible to visualize the effects of cytokinins on plant growth and morphogenesis as the ultimate expression of changes in the relative rates of protein biosynthesis brought about by the hormonal control of the synthesis and function of particular tRNA species." The influence of free cytokinins on RNA and protein synthesis should also be considered as a possible mechanism for cytokinin regulation (Sections III,D and V).

V. Hormone-Modified RNA Synthesis by Isolated Nuclei and Chromatin

The early *in vitro* studies directed to an analysis of the influence of plant hormones on RNA synthesis with isolated nuclei usually gave variable results. No effect of the hormone was obtained or the results were often suspect because of bacterial contamination. Since plant tissues have rigid cell walls, the isolation of intact nuclei also presented a major problem. A further complication was the possibility that a variable loss of nuclear material during isolation might be associated with the fact that an occasional experiment would show a hormone enhancement of RNA synthesis when the next several would not. That this was, in fact, the case was indicated by experiments of Johri and Varner (1968). They showed that nuclei isolated from light-grown dwarf pea synthesized RNA at some basal rate. This rate of RNA synthesis was not altered by the addition of GA to the incubation medium. However, when GA was included in the extraction medium, and enhanced rate of RNA synthesis was observed. Relative to the hormonal nature of this response, Johri and Varner showed that 10^{-8} M GA_3 (a gibberellin which elicits a growth response in the dwarf pea) was sufficient to elicit the maximum response; another gibberellin, GA_8, which does not cause a growth response in this system did not alter the rate of RNA synthesis.

Nearest neighbor analysis of the RNA product from control and GA-treated nuclei showed that GA caused a qualitative difference in the RNA being synthesized as well as an increase in the level of RNA synthesis. Johri and Varner suggested (1) that the GA effect on RNA synthesis could be due to an increase in template sites on DNA and/or the result of an increase in the activity of RNA polymerase and (2) that some factor(s) present in the nucleus or cytoplasm is involved before GA causes the final enhancement of RNA synthesis. GA, when included during isolation of nuclei, could prevent the loss of factor from the nucleus or conceivably be involved in transporting a cytoplasmic factor into the nucleus.

Additional evidence for hormone–factor interactions in the enhancement of RNA synthesis by isolated nuclei comes from experiments of Matthysse and Phillips (1969). They showed that nuclei-rich preparations from tobacco and soybean culture cells and from pea buds responded to auxin (2,4-D) by a 50 to 120% increase in the rate of RNA synthesis (Table I). However, the response was obtained with tobacco and soybean nuclei only when isolated in the presence of auxin. When these nuclei which were isolated in the presence of auxin were washed once by sedimentation from the incubation medium in the absence of auxin, they no longer showed enhanced RNA synthesis in response to auxin. This appeared to be due to the loss of some required factor from the nuclei, since the addition of the supernatant wash back to the washed nuclei restored the sensitivity to auxin (Table I). The factor was purified somewhat by gel filtration and shown to be heat labile and presumably

TABLE I

INTERACTION OF AUXIN AND "FACTOR" IN THE ENHANCEMENT
OF RNA SYNTHESIS BY ISOLATED NUCLEI AND CHROMATIN[a]

Isolation medium	Incubation medium	RNA synthesis (% of control)
	Nuclei	
− Auxin	− Auxin	100
− Auxin	+ Auxin	98 ± 7
+ Auxin	+ Auxin	170 ± 20
− Auxin	+ Factor	110 ± 30
− Auxin	+ Auxin + factor	170 ± 25
− Auxin	+ Auxin + boiled factor	90 ± 20
	Isolated chromatin	
	− Auxin	100
	+ Auxin	86 ± 4
	+ Factor	100
	+ Auxin + factor	170 ± 25

[a] Data adapted from Matthysse and Phillips (1969).

a protein. The factor or auxin alone had no effect on RNA synthesis by pea bud chromatin fortified with *E. coli* RNA polymerase, but when added together a marked enhancement of RNA synthesis was observed. Experiments using saturating levels of *E. coli* RNA polymerase indicated that the effect of auxin and the "hormone-reactive protein" was on chromatin template availability rather than directly on RNA polymerase.

Subsequently, Matthysse and Abrams (1970) also showed a nuclei–factor–cytokinin response at the level of RNA synthesis, similar to that discussed above for auxin.

Another approach which has been used in the study of the influence of auxin on RNA synthesis has been the isolation of chromatin from control and hormone-treated seedlings (e.g., O'Brien *et al.*, 1968; Holm *et al.*, 1970; Johnson and Purves, 1970). Such chromatin from auxin-treated soybean hypocotyl shows a severalfold enhanced synthesis of RNA relative to control chromatin (Fig. 15). A marked auxin response at the chromatin level is apparent within 4 hr after seedlings are treated (O'Brien *et al.*, 1968). The addition of *E. coli* RNA polymerase to chromatin isolated from control and auxin-treated plants markedly increases the rate of RNA synthesis (Fig. 15). Both chromatin preparations saturated at the same level of added *E. coli* RNA polymerase (but the difference in the amount of RNA synthesis persisted). O'Brien *et al.* 1968) interpreted these results to mean that auxin treatment causes an increase in the level

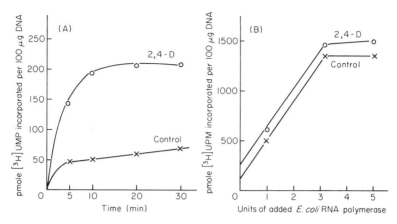

FIG. 15. (A) The kinetics of RNA synthesis directed by chromatin, and its associated RNA polymerase, extracted from the hypocotyl of control and 2,4-D-treated soybean seedlings. Seedlings were sprayed with 2,4-D (1000 μg/ml) 12 hr prior to chromatin isolation. (B) Saturation of chromatin DNA from control and 2,4-D-treated soybean seedlings with exogenous *E. coli* RNA polymerase. Chromatin DNA (0.3–0.4 μg DNA) and its associated RNA polymerase from control and 2,4-D-treated tissue was first assayed to determine the endogenous RNA polymerase activity. These levels of activity were plotted as the initial points on the Y axis of the saturation curves. In separate reaction mixtures *E. coli* RNA polymerase (1, 3, and 5 units, respectively) was added to obtain the remaining points on the saturation curves. (From O'Brien *et al.*, 1968.)

of chromatin-associated RNA polymerase rather than modification in template availability. While there are insufficient data at present to make a definitive conclusion about the role of auxin in enhancing chromatin-directed RNA synthesis, there are some results which indicate that auxin must alter the template availability. First, the auxin response in terms of amount of RNA synthesis in the absence of added polymerase is maintained in the presence of saturating levels of *E. coli* RNA polymerase (Fig. 15 and Holm and Key, 1971). Second, the RNA product from control and auxin chromatin has a different nearest neighbor frequency (Holm *et al.*, 1970). This difference could result, however, from a differential effect of auxin on different RNA polymerases which function to synthesize different types of RNA (Roeder and Rutter, 1969). (RNA polymerases will be discussed briefly in Section VI.) Finally, as discussed in Section III,A, Thompson and Cleland (1971) have shown that treatment of intact seedlings with auxin leads to changes in the RNA sequences being transcribed (Fig. 3). In other studies directed to an understanding of the chromatin response, Hardin *et al.* (1970) isolated a "factor" from control seedlings which causes an increase in RNA synthesis by control chromatin but not by chromatin isolated from auxin-treated tissue. Their working hypothesis is that auxin has caused the association of this factor with the chromatin in auxin-treated seedlings.

Venis (1971) recently published some most interesting results which may relate to the observations discussed above. In an elegant approach to the study of auxin-factor (or receptor) interaction, Venis coupled a 2,4-D–lysine derivative to agarose. He then compared the binding of protein from pea stem tissue to agarose and 2,4-D-substituted agarose columns. There was a small amount of protein bound to the auxin-modified agarose column in starting buffer while none was bound to the agarose. Probably because of the low specific activity auxin available, Venis has been unable to demonstrate binding by equilibrium dialysis of labeled auxin to the protein fraction. However, when added to an RNA synthesizing medium containing DNA and E. coli RNA polymerase, this protein fraction enhances RNA synthesis by 40 to 200% in different preparations. This "factor" enhancement of RNA synthesis does not depend on addition of auxin to the reaction. Venis obtained suggestive evidence that the "factor" may be acting by permitting initiation at regions of DNA not otherwise transcribed or by facilitating chain release and reinitiation at the same points. While there is no demonstrated interaction, either physically or biologically, of auxin and factor after isolation, it is certainly significant that the fraction was isolated on an auxin-derived absorbent. Such a fraction could be analogous to the "factor" studied by Hardin et al. (1970).

Possibly of importance in the regulation of chromatin activity by auxin are the observations of Venis (1968) and of Yasuda and Yamada (1970) which show association of auxin with histones. Using equilibrium dialysis Venis showed an association between 2,4-D and histones presumably by ionic interaction. In their work Yasuda and Yamada demonstrated a covalent interaction of 2,4-D with histones. Treatment with 2,4-D caused an apparent decrease in the level of histone and an increase in the level of acidic protein. These results, however, were due to the basicity of histones being changed (to acidic proteins) upon complexing with auxin. They further suggested that this interaction related to callus induction by 2,4-D.

VI. Possible Model Systems for Developmental Regulation at the Level of Transcription

Hypothetical mechanisms or model systems by which hormones might alter the transcriptional process, and thus the developmental pattern of an organism, must include consideration of the functioning of RNA polymerase. Three systems, phage infection of *E. coli,* sporulation in *B. subtilis,* and the transcription of catabolite-sensitive genes in bacteria, which have in common differential gene transcription, will be discussed here as possible elementary model systems for developmental regulation in eukaryotes.

Although RNA polymerase has been studied for a number of years in both prokaryotic and eukaryotic systems, the complexity of the enzyme is only now being realized (Burgess, 1971). The bacterial polymerase is composed of a multisubunit "core enzyme" and a sigma subunit. The sigma factor of the complete polymerase is apparently required for specificity of initiation (e.g., Travers and Burgess, 1969; Bautz and Bautz, 1970; Suigura *et al.,* 1970), whereas the core enzyme alone "initiates" randomly with purified DNA and gives low rates of synthesis of random size RNA molecules. An additional factor, psi, interacts with the complete RNA polymerase of bacteria to elicit the transcription of rRNA (Travers *et al.,* 1970).

Eukaryotic systems are now known to contain multiple RNA polymerases (e.g., Roeder and Rutter, 1969, 1970; Horgen and Griffin, 1971; Mondal *et al.,* 1970; Cold Spring Harbor Symposium, 1970). Of the two major RNA polymerases, one is localized in the nucleoplasm while the other is in the nucleolus (Roeder and Rutter, 1970). Although the exact structure of the eukaryotic RNA polymerase(s) is not known, they appear to be composed of multisubunits similar to the bacterial enzyme. To date, however, little is known about these RNA polymerases relative to regulatory elements (e.g., sigma and psi factors).

The "developmental events," phage infection and sporulation in bacteria, are associated with changes in the sigma factor and/or core–subunit makeup of RNA polymerase. The endogenous *E. coli* RNA polymerase (complete) specifically and selectively transcribes "preearly" genes of T_4 phage (Bautz *et al.,* 1969). On the other hand, the T_4 factor (which was isolated from T_4-infected cells and which is functionally analogous to sigma factor) together with the "core" enzyme directs the transcription of other regions of the T_4 genome (Travers, 1969, 1970). Similar to phage infection of *E. coli,* changes in RNA polymerase have been implicated in the control of sporulation in *B. subtilis.* Early in the process leading to sporulation, *B. subtilis* RNA polymerase changes dramatically in template specificity (Sonenshein and Losick, 1970). This change in specificity relates to a change in subunit structure of the RNA polymerase from sporulating cells relative to vegetative cells (Sonenshein and Losick, 1970; Losick *et al.,* 1970). These changes in the structural makeup of RNA polymerase are postulated to control the "turn off" of host genes in the case of phage infection or vegetative genes in the case of sporulation and to "turn on" additional phage genes and sporulation-specific genes.

Another possibly more important model for developmental regulation by small molecules such as hormones comes from work on the positive control of the transcription of catabolite-sensitive genes of bacteria. For many years it was known that *E. coli* fermenting glucose made less inducible enzymes than organisms fermenting other substrates such as lactate or acetate (glucose repression or catabolite repression). That cyclic AMP might be involved in this process was indicated by the work of Markman and Sutherland (1965) and by several other studies reviewed by Pastan and Perlman (1971). Recent work from the laboratories of Zubay *et al.* (1970) and Pastan and Perlman (1970, 1971) shows that cyclic AMP in concert with a protein factor (cyclic AMP receptor protein, or CRP) does, in fact, serve as a positive control system for the transcription of catabolite-sensitive genes. Cyclic AMP and CRP are required for RNA polymerase binding to the *lac* promoter region of the *lac* DNA and formation of the RNA polymerase–DNA initiation complex. In this case, the *in vitro* transcription of the *lac* operon into *lac* mRNA then requires cyclic AMP and CRP along with RNA polymerase and DNA containing the *lac* genes.

These studies taken together with those on hormone–factor interactions described in Section V should serve as working models in experiments designed to elucidate the mechanism(s) of hormone-enhanced RNA synthesis.

VII. Summary

The evidence presented above generally supports the view that at least some hormonal responses are mediated via RNA and protein synthesis (e.g., gibberellin regulation of seed germination, auxin control of cell elongation, and ethylene-mediated abscission). This is not to say, however, that the "site of action" is directly on either transcription- or translation-related processes. The fast growth responses (1 to 3 min) to auxin, while not auxin-specific, probably are not mediated initially via effects on either RNA or protein synthesis. On the other hand, the *in vitro* hormone-induced enhancement of RNA synthesis by nuclei and the hormone–"factor"-mediated enhancement of chromatin-directed RNA synthesis are indeed suggestive of a role for hormones on the transcription process. The work with abscisic acid, both in the barley system discussed here and in the cotton system discussed by Dure in Chapter 2, indicates that this hormone is involved in the control of translation of specific mRNA's associated with the germination process. The occurrence of cyto-kinins in tRNA and the apparent significance of this base in the function-ing of tRNA are again suggestive of an involvement of cytokinins at some level in the control of protein synthesis. Thus while considerable progress has been made in relating hormone function to the control of RNA and protein synthesis, much more remains to be learned.

The relationship of the fast response to auxin to the requirement for RNA and protein synthesis in the auxin control of cell elongation needs to be investigated in detail. While there is little if any definitive evidence for the occurrence of cyclic AMP and the associated adenylcyclase and phosphodiesterase in plant systems, the possibility should be considered (based on the vast accumulating literature on the significance of cyclic AMP as a secondary messenger in many hormone responses in animals, e.g., Pastan and Perlman, 1971; Jost and Rickenberg, 1971) that a plant hormone might regulate very diverse physiological processes via cyclic AMP-mediated protein kinases. As an example, auxin might interact at the cell membrane to activate cyclic AMP formation. This in turn might lead to the activation of a cell membrane-mediated hydrogen pump (see Hager et al., 1971) to cause the fast growth response which can be mediated both by a pH drop and by auxin. One could visualize the hypothetical auxin-induced increase in cyclic AMP as leading to the activa-tion of cell wall biosynthetic enzymes and an increase in cell wall bio-synthesis. The cyclic AMP could also be involved in phosphorylation of chromatin proteins (e.g., Langan, 1969), and through this or some other mechanism lead to an enhancement of RNA synthesis (e.g., Dokas and Kleinsmith, 1971). This suggestion, of course, only represents an attempt to relate three seemingly diverse physiological responses to auxin to a common "site of action." Auxin does enhance these processes and there is at least one very preliminary report that auxin may enhance the cyclic AMP level of Bengal gram seedlings (Azhar and Murti, 1971). (This and unpublished data from other laboratories provide only suggestive evidence for the occurrence of cyclic AMP in plants.) There are also reports of effects of exogenously added cyclic AMP on hydrolase synthe-sis (normally GA-mediated) in seeds (Galskey and Lippincott, 1969; Duffus and Duffus, 1969; Pollard, 1970). Thus the search for cyclic AMP and related enzymes is a potentially fruitful area of investigation and likely would prove enlightening relative to hormonal regulation in plants.

The significance of cytokinins in tRNA and the relationship to the hormonal function of free cytokinins need to be established. Is the pres-ence of the iPA or 2 methylthio-iPA adjacent to the anticodon essential to the functioning of the aminoacyl-tRNA in protein synthesis, or does it possibly relate to some "regulatory" function of aminoacyl-tRNA? Do

free cytokinins function in some way to regulate the attachment of the
isopentenyl (or one of the other constituents) group to tRNA (e.g., via
affecting precursor availability or by modulating the activity of the attachment enzyme)? The availability of anticytokinins (Hecht *et al.*,
1971) should be valuable in answering these and other questions relating
to cytokinin action.

While there are no definitive answers on how plant hormones function
to regulate the many physiological responses which they affect, considerable progress has been made. New approaches based on current research
in both prokaryotic and eukaryotic organisms should make for considerable progress in understanding the role of plant hormones in the control
of transcriptional and translational processes, whether by some direct
interaction of the hormone with some component of the regulatory system
or via the action of a secondary "messenger" (e.g., cyclic AMP).

Supplementary Readings

Galston, A. W., and Davies, P. J. (1969). *Science* **163**, 1288.
Galston, A. W., and Davies, P. J. (1970). "Control Mechanisms in Plant Development." Prentice-Hall, Englewood Cliffs, New Jersey.

References

Abeles, F. B., (1969). *Plant Physiol.* **44**, 447.
Abeles, F. B., and Holm, R. E. (1966). *Plant Physiol.* **41**, 1337.
Anderson, J., and Rowan, K. (1966). *Biochem. J.* **101**, 15.
Anderson, J. M. (1972). Ph.D. Dissertation, Purdue University, Lafayette, Indiana.
Anderson, M. B., and Cherry, J. H. (1969). *Proc. Nat. Acad. Sci. U.S.* **62**, 202.
Azhar, S., and Murti, C. R. K. (1971). *Biochem. Biophys. Res. Commun.* **43**,
    58.
Balz, H. P. (1966). *Planta* **70**, 207.
Barkley, G. M., and Evans, M. (1970). *Plant Physiol.* **45**, 143.
Bautz, E. K. F., and Bautz, F. A. (1970). *Nature (London)* **226**, 1219.
Bautz, E. K. F., Bautz, A. F., and Dunn, J. J. (1969). *Nature (London)* **223**,
    1022.
Beevers, L. (1966). *Plant Physiol.* **41**, 1074.
Brandes, H., and Kende, H. (1968). *Plant Physiol.* **43**, 827.
Broughton, W. J. (1968). *Biochim. Biophys. Acta* **155**, 308.
Brown, H., and Escombe, F. (1898). *Proc. Roy. Soc., London* **63**, 3.
Burgess, R. R. (1971). *Annu. Rev. Biochem.* **40**, 711.
Burrows, W. J., Armstrong, D. J., Skoog, F., Hecht, S. M., Dammam, L. G., Leonard,
    N. J., and Occolowitz, J. (1970). *Biochemistry* **9**, 1867.
Burrows, W. J., Skoog, F., and Leonard, N. J. (1971). *Biochemistry* **10**, 2189.
Chandra, G. R., and Varner, J. E. (1965). *Biochim. Biophys. Acta* **108**, 583.

Chrispeels, M., and Hanson, J. B. (1962). *Weeds* **10**, 123.

Chrispeels, M., and Varner, J. E. (1966). *Nature (London)* **212**, 1066.

Chrispeels, M., and Varner, J. E. (1967a). *Plant Physiol.* **42**, 398.

Chrispeels, M., and Varner, J. E. (1967b). *Plant Physiol.* **42**, 1008.

Cleland, R. (1971a). *Annu. Rev. Plant Physiol.* **22**, 197.

Cleland, R. (1971b). *Planta* **99**, 1.

de Crombrugghe, B., Chen, B., Anderson, W., Nissley, P., Gottesman, M., Pastan, I., and Perlman, R. (1971). *Nature (London), New Biol.* **231**, 139.

de Hertogh, A. A., McCune, D. C., Brown, J., and Antonie, D. (1965). *Contrib. Boyce Thompson Inst.* **23**, 23.

Dokas, L. A., and Kleinsmith, L. J. (1971). *Science* **172**, 1237.

Duffus, C. M., and Duffus, J. H. (1969). *Experientia* **25**, 381.

Eb, A. A., and Nieuwdorp, P. J. (1967). *Acta Bot. Neer.* **15**, 690.

Evans, M. L., and Ray, P. M., (1969). *J. Gen. Physiol.* **53**, 1.

Evans, M. L., Ray, P. M., and Reinhold, L. (1971). *Plant Physiol.* **47**, 335.

Evins, W. H., and Varner, J. E. (1971). *Proc. Nat. Acad. Sci. U.S.* **68**, 1631.

Fan, Der-Fong, and Maclachlan, G. A. (1967). *Plant Physiol.* **42**, 1114.

Ferrari, T. E., and Varner, J. E. (1969). *Plant Physiol.* **44**, 85.

Filner, P., and Varner, J. (1967). *Proc. Nat. Acad. Sci. U.S.* **58**, 1520.

Fittler, F., and Hall, R. (1966). *Biochem. Biophys. Res. Commun.* **25**, 441.

Fletcher, R. A., and Osborne, D. J. (1966). *Can. J. Bot.* **44**, 739.

Fox, J. E. (1966). *Plant Physiol.* **41**, 75.

Fox, J. E., and Chen, C.-M. (1967). *J. Biol. Chem.* **242**, 4490.

Galskey, A. G., and Lippincott, J. A. (1969). *Plant Cell Physiol.* **10**, 607.

Gefter, M., and Russell, R. (1969). *J. Mol. Biol.* **39**, 145.

Haberlandt, G. (1890). *Ber. Deut. Bot. Ges.* **8**, 40.

Hager, A. H., Menzel, H., and Kraus, A. (1971). *Planta* **100**, 47.

Hardin, J. W., O'Brien, T. J., and Cherry, J. H. (1970). *Biochim. Biophys. Acta* **224**, 667.

Hayashi, H., Fisher, H., and Soll, D. (1969). *Biochemistry* **8**, 3680.

Hecht, S. M., Leonard, N. J., Burrows, W. J., Armstrong, D. J., and Occolowitz, F. (1969). *Science* **166**, 1272.

Hecht, S. M., Bock, R. M., Schmitz, R. Y., Skoog, F., and Leonard, N. J. (1971). *Proc. Nat. Acad. Sci. U.S.* **68**, 2608.

Helgeson, J. P. (1968). *Science* **161**, 974.

Holm, R. E., and Key, J. L. (1969). *Plant Physiol.* **44**, 1295.

Holm, R. E., and Key, J. L. (1971). *Plant Physiol.* **47**, 606.

Holm, R. E., O'Brien, T. J., Key, J. L., and Cherry, J. H. (1970). *Plant Physiol.* **45**, 41.

Horgen, P. A., and Griffin, D. H. (1971). *Proc. Nat. Acad. Sci. U.S.* **68**, 338.

Ingle, J., Holm, R. E., and Key, J. L. (1965). *J. Mol. Biol.* **11**, 730.

Jacobson, J., and Varner, J. E. (1967). *Plant Physiol.* **42**, 1596.

Johnson, K. D., and Purves, W. K. (1970). *Plant Physiol.* **46**, 581.

Johri, M., and Varner, J. E. (1968). *Proc. Nat. Acad. Sci. U.S.* **59**, 269.

Jones, R. L. (1969a). *Plant Physiol.* **44**, 101.

Jones, R. L. (1969b). *Planta* **87**, 119.

Jones, R. L. (1969c). *Planta* **88**, 73.

Jones, R. L. (1971). *Plant Physiol.* **47**, 412.

Jost, J. P., and Rickenberg, H. V. (1971). *Annu. Rev. Biochem.* **40**, 741.

Kende, H., and Tavares, J. (1968). *Plant Physiol.* **43**, 1244.

Key, J. L. (1964). *Plant Physiol.* **39**, 365.

Key, J. L. (1969). *Annu. Rev. Plant Physiol.* **20**, 449.

Key, J. L., and Hanson, J. B. (1961). *Plant Physiol.* **36**, 145.

Key, J. L., and Ingle, J. (1964). *Proc. Nat. Acad. Sci. U.S.* **52**, 1382.

Key, J. L., and Ingle, J. (1968). *In* "Biochemistry and Physiology of Plant Growth Substances," p. 711. Runge Press, Ottawa.

Key, J. L., and Shannon, J. C. (1964). *Plant Physiol.* **39**, 360.

Key, J. L., Lin, C. Y., Gifford, E. M., Jr., and Dengler, R. (1966). *Bot. Gaz. (Chicago)* **127**, 87.

Key, J. L., Barnett, N. M., and Lin, C. Y. (1967). *Ann. N.Y. Acad. Sci.* **144**, 49.

Key, J. L., Leaver, C. J., Cowles, J. R., and Anderson, J. M. (1972). *Plant Physiol.* **49**, 783.

Khan, A. A., (1971). *Science* **171**, 853.

Kuraishi, S. (1968). *Physiol. Plant.* **21**, 78.

Langan, T. A. (1969). *Proc. Nat. Acad. Sci. U.S.* **64**, 1276.

Lewis, L. N., and Varner, J. E. (1970). *Plant Physiol.* **46**, 194.

Lin, C. Y., and Key, J. L. (1968). *Plant Cell Physiol.* **9**, 553.

Litwack, M. D., and Peterkofsky, A. (1971). *Biochemistry* **10**, 994.

Losick, R., and Shorenstein, R. G. (1970). *Nature (London)* **227**, 910.

MacLeod, A. M. (1966). *J. Inst. Brew., London* **72**, 580.

Madison, J. T., and Kung, H.-K. (1967). *J. Biol. Chem.* **242**, 1324.

Markman, R. S., and Sutherland, E. W. (1965). *J. Biol. Chem.* **240**, 1309.

Masuda, Y. (1959). *Physiol. Plant.* **12**, 324.

Masuda, Y., Tanimoto, E., and Wada, S. (1967). *Physiol. Plant.* **20**, 713.

Matthysse, A. G., and Abrams, M. (1970). *Biochim. Biophys. Acta* **199**, 511.

Matthysse, A. G., and Phillips, C. (1969). *Proc. Nat. Acad. Sci. U.S.* **63**, 897.

Mondal, H., Mandal, R. K., and Biswas, B. B. (1970). *Biochem. Biophys. Res. Commun.* **40**, 1194.

Morris, R. O. (1966). *Biochim. Biophys. Acta* **127**, 273.

Nooden, L. D., and Thimann, K. V. (1964). *Proc. Nat. Acad. Sci. U.S.* **50**, 194.

O'Brien, T. J., Jarvis, B., Cherry, J. H., and Hanson, J. B. (1968). *Biochim. Biophys. Acta* **169**, 35.

Osborne, D. J. (1965). *J. Sci. Food Agr.* **16**, 1.

Paleg, L. (1960). *Plant Physiol.* **35**, 293.

Pastan, I., and Perlman, R. L. (1970). *Science* **169**, 339.

Pastan, I., and Perlman, R. L. (1971). *Nature (London), New Biol.* **229**, 5.

Penny, P., and Galston, A. W. (1966). *Amer. J. Bot.* **53**, 1.

Pollard, C. J. (1970). *Biochim. Biophys. Acta* **201**, 511.

Ray, P. M. (1969). *Develop. Biol., Suppl.* **3**, 172.

Ray, P. M., and Ruesink, A. W. (1962). *Develop. Biol.* **4**, 377.

Rayle, D. L., and Cleland, R. (1970). *Plant Physiol.* **46**, 250.

Rayle, D. L., Evans, M. L., and Hertel, R. (1970a). *Proc. Nat. Acad. Sci. U.S.* **65**, 184.

Rayle, D. L., Haughton, P. M., and Cleland, R. (1970b). *Proc. Nat. Acad. Sci. U.S.* **67**, 1814.

Rebstock, T. L., Hamner, C. L., and Sell, H. M. (1954). *Plant Physiol.* **29**, 490.

Richmond, A., and Lang, A. (1957). *Science* **125**, 650.

Roeder, R. G., and Rutter, W. J. (1969). *Nature (London)* **224**, 235.

Roeder, R. G., and Rutter, W. J. (1970). *Proc. Nat. Acad. Sci. U.S.* **65**, 675.

Sacher, J. A. (1967). *Exp. Gerontol.* **2**, 261.

Shannon, J. C., Hanson, J. B., and Wilson, C. M. (1964). *Plant Physiol.* **39**, 804.

Skoog, F. (1954). *Brookhaven Symp. Biol.* **6**, 1.

Skoog, F., and Armstrong, D. J. (1970). *Annu. Rev. Plant Physiol.* **21**, 359.

Sonenshein, A. L., and Losick, R. (1970). *Nature (London)* **227**, 906.

Srivastava, B. (1968). *Biochem. Biophys. Res. Commun.* **32**, 533.

Srivastava, B., and Ware, G. (1965). *Plant Physiol.* **40**, 62.

Stafford, H. A., and Galston, A. W. (1970). *Plant Physiol.* **46**, 763.

Steward, F. C., and Krikorian, A. D. (1971). "Plants, Chemicals and Growth." Academic Press, New York.

Suigura, M., Okamoto, T., and Takanami, M. (1970). *Nature (London)* **225**, 598.

Tavares, J., and Kende, H. (1970). *Phytochemistry* **9**, 1763.

Thompson, W. F., and Cleland, R. E. (1971). *Plant Physiol.* (in press).

Travers, A. A. (1969). *Nature (London)* **223**, 1107.

Travers, A. A. (1970). *Nature (London)* **225**, 1009.

Travers, A. A., and Burgess, R. R. (1969). *Nature (London)* **222**, 538.

Travers, A. A., Kamen, R. I., and Schleif, R. F. (1970). *Nature (London)* **228**, 748.

Trewavas, A. (1968). *Phytochemistry* **7**, 673.

Trewavas, A. (1968). *Progr. Phytochem* **1**, 113.

van Overbeek, J. (1966). *Science* **152**, 721.

van Overbeek, J. (1968). *Sci. Amer.* **219**, 75.

Varner, J. E. (1964). *Plant Physiol.* **39**, 413.

Varner, J. E., and G. R. Chandra (1964). *Proc. Nat. Acad. Sci. U.S.* **52**, 100.

Varner, J. E., Chandra, G. R., and Chrispeels, M. J. (1965). *J. Cell. Comp. Physiol.* **66**, 55.

Venis, M. A. (1964). *Nature (London)* **202**, 900.

Venis, M. A. (1968). *In* "Biochemistry and Physiology of Plant Growth Substances," p. 761. Runge Press, Ottawa.

Venis, M. A. (1971). *Proc. Nat. Acad. Sci. U.S.* **68**, 1824.

von Abrams, G. J., and Pratt, H. K. (1968). *Plant Physiol.* **43**, 1271.

Wollgiehn, R. (1967). *Soc. Exp. Biol.* **21**, 231.

Yasuda, T., and Yamada, Y. (1970). *Biochem. Biophys. Res. Commun.* **40**, 649.

Yomo, H. (1961). *Chem. Abstr.* **55**, 26145.

Zachau, H., Dutting, D., and Feldman, H. (1966). *Angew. Chem.* **78**, 392.

Zubay, G., Schwartz, D., and Beckwith, J. (1970). *Proc. Nat. Acad. Sci. U.S.* **66**, 104.

4

Transitions in Differentiation by the Cellular Slime Molds*

JAMES H. GREGG and W. SUE BADMAN

I. Introduction

A mature fruiting body of a cellular slime mold is composed of a slender stalk which supports a mass of spores (Figs. 1–9). On germination of the spores, one amoeboid cell emerges from each spore and proceeds to undergo binary fission, and as a consequence of cell division large numbers of independent myxamoebae appear in a culture. Subsequently the myxamoebae aggregate to form colonies or pseudoplasmodia, each of which is composed of thousands of cells. The migrating pseudoplasmodia are composed of two types of cells which may be distinguished by appro-

* This investigation was supported in part by Public Health Service Research Career Programs Award 5-K3-HD-15,780 and Research Grant GM-10138 from the National Institutes of Health.

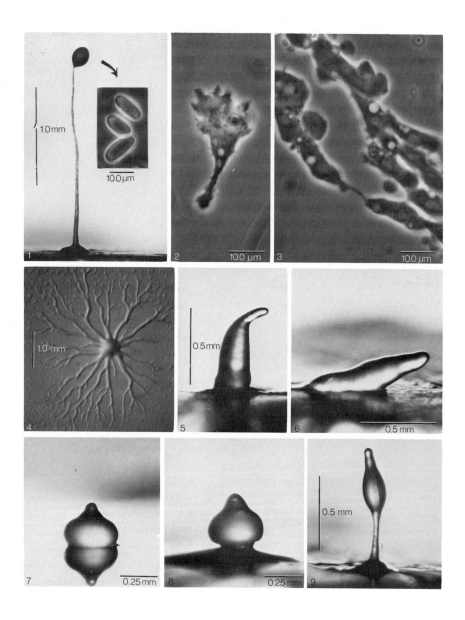

Figs. 1–9. *Dictyostelium discoideum.* From Gregg (1966). Fig. 1. Mature sorocarp and spores. Fig. 2. Vegetative myxamoeba. Fig. 3. Myxamoebae removed from aggregating streams. Fig. 4. Pattern of aggregation. Fig. 5. Late aggregate. Fig. 6. Migrating pseudoplasmodium. Fig. 7. Preculmination. Figs. 8 and 9. Culmination.

priate methods. These two cell types, the prestalks and prespores, eventually differentiate into mature stalk and spore cells. Thus, fruiting body formation depends on the gradual transition of a common cell type to specific types during the morphogenetic process.

Myxamoebae which are prevented from entering an aggregate are incapable of differentiating into mature stalks or spores. Therefore, cell association beginning at aggregation appears to be necessary in ordering the course of differentiation of the cells. Consequently this review of development in the cellular slime molds will emphasize the factors concerned with the divergence of a common cell type to specific types and the role of cell association in this process.

II. Variability among the Myxamoebae

A. VEGETATIVE STAGE

There is no evidence available which reveals that vegetative myxamoebae vary in their physiological or biochemical properties. They are identical in appearance with phase microscopy and are indistinguishable at the ultrastructural level (Gregg and Badman, 1970). Thus vegetative myxamoebae are considered to be homogeneous. However, cell homogeneity does not persist as the preaggregation stage is approached.

B. PREAGGREGATION STAGE

Takeuchi (1963) first revealed that two types of preaggregating myxamoebae could be distinguished by the variations in the numbers of cytoplasmic granules which were available to combine with fluorescent antibody. The degree of staining with fluorescent antibody resulted in cells which were brightly stained or remained relatively dark. Subsequently, Takeuchi (1969) demonstrated that two major classes of preaggregating myxamoebae could be isolated by virtue of their density differences. These heavy and light classes of myxamoebae were separated by centrifugation in a dextrin solution (Fig. 10). The density variations were attributed to either unequal cell division or growth rate differences during the vegetative stage. It seems plausible that cells growing nonsynchronously might vary in their densities depending on the unique ratios of substrates synthesized or utilized by the individuals. In addition, the bacterial residues retained during growth could contribute to variable densities among a population of myxamoebae.

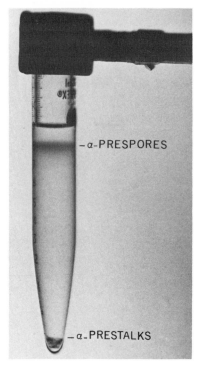

FIG. 10. *D. discoideum* preaggregation myxamoebae centrifuged in dextrin solution reveals the separation of the light (α-prespores) and heavy (α-prestalks) classes of myxamoebae.

III. Pseudoplasmodium Formation

A. AGGREGATION STAGE

Aggregation of the myxamoebae is, of course, the necessary prelude to the formation of a multicellular fruiting body. A chemotactic substance originally named acrasin by Bonner (1947) is responsible for initiating the morphogenetic movements of aggregation. Subsequently an active chemotactic factor in acrasin proved to be cyclic adenosine 3',5'-phosphate (Konijn *et al.*, 1967) (Fig. 11). Under the influence of cyclic AMP, or acrasin, which is actively secreted by small centers or groups of cells, the myxamoebae of certain species are oriented toward a center resulting in the formation of the multicellular pseudoplasmodium.

In responding to the acrasin, the myxamoebae become adhesive (Shaffer, 1958), a property which undoubtedly promotes the integrity of the pseudoplasmodium during morphogenesis. Of particular interest

FIG. 11. Acrasia, the witch in Edmund Spenser's *Faerie Queene* was engaged in pursuits which suggested that "acrasin" would be an appropriate name for the chemotactic substance secreted by the *Acrasiales* (Bonner, 1949a). Twenty years of probing induced Acrasia to reveal a major ingredient in her recipe as cyclic 3′,5′-AMP (Konijn *et al.*, 1967).

is the fact that the anterior prestalk cells which secrete acrasin in greatest quantity in the slug (Bonner, 1949b) also appear to maintain relatively greater adhesiveness than the prespore cells. This observation is based upon electron micrographic studies which reveal that prestalk cell surfaces are closely associated, while prespore cells exhibit a more casual type of contact (Maeda and Takeuchi, 1969). The adhesive properties acquired at aggregation have been attributed to the appearance of new surface antigens by Gregg (1956) and Sonneborn *et al.* (1964) and are considered to be polysaccharide in nature by Gerisch (1969). Subsequently Beug *et al.* (1970) prepared univalent antiparticulate and anticarbohydrate sera from aggregation-competent *Dictyostelium discoideum* myxamoebae. Although the antiparticulate serum inhibited cell contacts among aggregating myxamoebae, the anticarbohydrate serum was not effective in this respect.

At least three antigens, possibly carbohydrate-associated proteins, compose the contact sites of aggregation-competent myxamoebae but the vegetative myxamoebae are almost devoid of these antigens. Vegetative myxamoebae are nonadhesive, which suggests that the antigens appearing at aggregation are concerned with the cell adhesion process.

B. Polarization of Pseudoplasmodium

Following aggregation, additional morphogenetic movements result in an elongated polarized slug shape which is assumed just prior to migration. The polarization of the cell mass is recognized by virtue of the fact that the direction of migration is dictated by only one end of the slug. This leading, or anterior end, is composed of prestalk cells, while prespore cells comprise the remainder. Aggregating myxamoebae are also polarized in that the nucleus is found predominantly in the anterior half of the cell (Bonner, 1950). The causal relationship, if any, between the polarization at the cellular and the pseudoplasmodial level is obscure.

However, the polarization process apparently does not depend on the orientating effect of acrasin as cell aggregates induced among vegetative myxamoebae in a shaker gained anterior–posterior polarity (Bonner, 1950). Therefore, polarity was attained by the cell mass in the absence of normal aggregation.

In *D. mucoroides* the differentiation of prestalk cells coincides with the initial elongation of the aggregate (Gregg, 1965). Bonner (1971) suggests that the prestalk cells, having greater mutual adhesiveness, are maintained at the anterior end of the cell mass. Consequently the unique elongate shape of the slug may be partially attributed to the existence of cells having qualitatively different adhesiveness. The factors dictating the form assumed by the pseudoplasmodium at various morphogenetic stages are undoubtedly manifold and worthy of investigation.

IV. The Migrating Pseudoplasmodium

A. Organization

Bonner (1959) determined by the use of vital dyes that a random mixing of the cells occurred in an aggregate. Thus the first cells that join the acrasin-secreting center do not necessarily become the anterior prestalk cells while the remainder of the cells assume a more posterior position. This discovery has further been substantiated by Takeuchi (1963, 1969). In this series of studies, Takeuchi determined that the position a particular class of preaggregating myxamoebae would eventually occupy in a mature migrating pseudoplasmodium could be predicted by certain properties of the class. Takeuchi noted that two major classes could be detected among the preaggregating myxamoebae. One class consists of cells which stain relatively poorly with fluorescent antisera, have a high specific gravity, and probably have a higher potential for becoming adhesive. The other class of myxamoebae may be defined as cells with the reciprocal values in those three categories.

A migrating pseudoplasmodium consists of two major cell types, the anterior prestalk cells and the posterior prespore cells. A variety of histochemical studies of slugs have revealed distinct differences between prestalk and prespore cells (Bonner *et al.*, 1955; Takeuchi, 1963; Gregg, 1965; Krivanek and Krivanek, 1958). Takeuchi (1963) and Gregg (1965) revealed that the anterior prestalk cells could readily be distinguished from the posterior prespore cells in *Dictyostelium* by the use of fluorescent antisera. The prespore cells which stained intensely were sharply contrasted with the prestalk cells which were very poorly stained (Fig. 12). Takeuchi (1963) observed that bright and dark cells existed among the preaggregating myxamoebae and suggested that the organization of the slug depended on a sorting of the two types of cells.

Additional evidence supporting the cell-sorting hypothesis was provided by Takeuchi (1969). This involved labeling the myxamoebae by culturing them in the presence of tritiated thymidine. By centrifuging the preaggregating cells in a density gradient Takeuchi isolated the fraction of labeled cells having the greatest specific gravity. The heavier labeled cells were then combined with a lighter nonlabeled fraction and

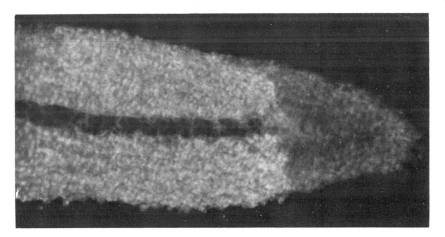

FIG. 12. *D. mucoroides* migrating pseudoplasmodium exhibiting the differential staining of the prespore and prestalk cells in the presence of fluorescent antiserum. Prespore cells consistently stain with greater intensity relative to the prestalk cells.

slug formation allowed to ensue. By autoradiography it was determined that the heavier cells composed the anterior prestalk end of the slug. Miller *et al.* (1969) have also confirmed that prestalk cells isolated from *D. discoideum* slugs have a relatively greater specific gravity than the prespore cells.

The evidence for cell sorting, however, is not in complete agreement. Farnsworth and Wolpert (1971) labeled *D. discoideum* myxamoebae with acridine orange absorbed to an ion exchange resin. Prestalk and prespore groups were grafted to slugs which continued to migrate. The failure of these grafts to regain their normal position led these authors to conclude that sorting out does not occur in a migrating slug. Francis and O'Day (1971) using a similar approach determined that cell sorting and redifferentiation is a continuous process during migration and is responsible for stabilizing prestalk–prespore proportions in the slug.

The existence of variability among the preaggregation myxamoebae suggests that sorting out is of necessity in slug formation. Although the specific type of variation observed among the myxamoebae previews their subsequent developmental fates, prestalk and prespore differentiation is not completed until the late aggregate is formed. This statement is made in view of the fact that the characteristic ultrastructural components found in prespore cells (Fig. 13) do not occur in the cells prior to late aggregation (Gregg and Badman, 1970) (Fig. 14). Under these circumstances it may be expedient to refer to the cells during this transition as α-prestalk cells and α-prespore cells until the late aggregate is completed.

B. REGULATION OF PROPORTIONS IN *Dictyostelium*

The size of an individual pseudoplasmodium depends on the number of cells entering the aggregate. Regardless of the ultimate size which may range from a few cells to thousands, the proportions of prestalk and prespore cells remain reasonably constant. In *D. discoideum* the prestalk cells constitute about 25% of the pseudoplasmodium while prespore cells compose the remainder (Bonner, 1957; Gregg and Bronsweig, 1956).

The sorting of α-prestalk and α-prespore cells is considered a significant factor in slug formation. However, proportionality in the slug cannot depend solely on the separation of these two cell types as this requires that the appropriate numbers of each cell type enter the aggregate. In order to establish proportionality under these circumstances, regulation must occur in the α-cell type in excess. The transition of the α-prestalk and α-prespore cells to prestalk and prespore cells occurs during the formation of the late aggregate (Gregg and Badman, 1970). Consequently the establishment of

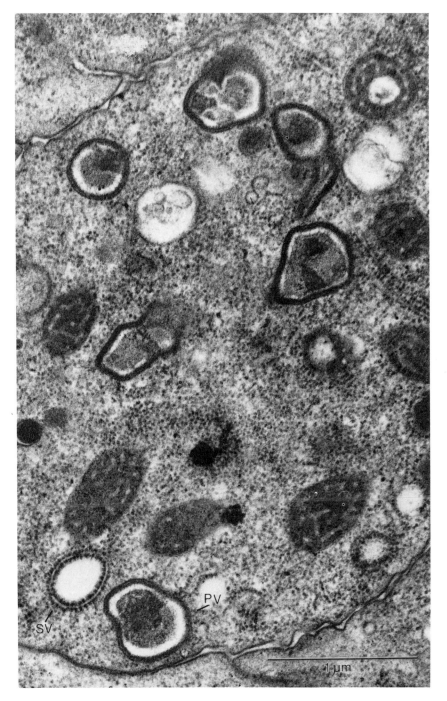

Fɪɢ. 13. Electron micrograph of a prespore cell revealing spore vesicles (SV) and prespore vacuoles (PV) neither of which exist in prestalk cells. ×40,500.

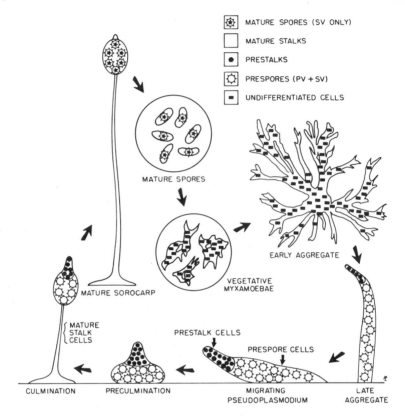

Fig. 14. Diagram of pattern of appearance, disappearance, and cellular distribution of spore vesicles (SV) and prespore vacuoles (PV) in *D. discoideum*. From Gregg and Badman (1970).

proportionality by regulation must be a concurrent event during this transitory period.

Dictyostelium mucoroides, unlike *D. discoideum*, produces a stalk during the entire migration stage. Despite the demand on the prestalk cells, their relative proportion remains essentially constant throughout migration (Fig. 15). This phenomenon depends on a consistent redifferentiation of prespore cells into prestalk cells. Thus, it appears that proportionality in *Dictyostelium* is achieved by cell interaction which in some obscure manner affects the differentiation of prestalk and prespore cells in the appropriate numbers.

It is also necessary to keep in mind that prestalk or prespore cell *groups* isolated from a mature migrating pseudoplasmodium are each capable of redifferentiating the missing cell type and again regulating their pro-

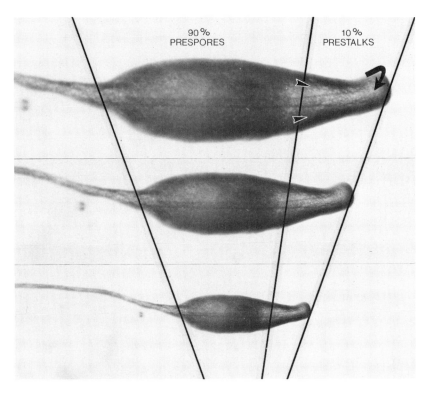

90%
PRESPORES

10%
PRESTALKS

FIG. 15. Illustration of the morphogenetic movements and pattern of cell differentiation which enables a *D. mucoroides* migrating pseudoplasmodium to maintain a constant proportionality between the prestalk and the prespore cells. The proportionality between the anterior and the posterior cell types which is established during slug formation ranges from approximately 5.0–17.0% prestalks to 83.0–95.0% prespores. The initial proportional relationship which depends upon the size of the slug is maintained throughout the migration phase despite the continual loss of prestalk cells in stalk differentiation. Arrows indicate the movement of prespore cells and their subsequent differentiation into mature stalk cells. Derived from Bonner (1957).

portions (Raper, 1940). Consequently, sorting out does not appear to be essential in the regulation of isolates.

V. Cell Association and Differentiation

A. REDIFFERENTIATION OF ASSOCIATED AND NONASSOCIATED MYXAMOEBAE

A major event in the development of the cellular slime molds consists of the formation of a fruiting body bearing mature spores. Aggregation

and cell association is considered to be essential in mature spore formation since the differentiation of isolated myxamoebae into spores has
never been observed.

The ability of myxamoebae to interact has been emphasized by a
variety of experiments involving morphogenetically deficient mutants.
Sussman and Lee (1955) determined that certain mutants were incapable
of completing development alone. However, if such mutants were combined, mature sorocarps were produced by synergistic action of the cells.
This synergistic response could only be achieved if the mutant cells were
physically associated and not when separated by thin agar membranes.

It has been established that α-prestalk and α-prespore cells originate
during the preaggregation stages (Takeuchi, 1963, 1969). By the late aggregation stage the α-prespores have completed the transition into
prespores as indicated by their characteristic ultrastructural components,
the prespore vacuoles (PV) (Hohl and Hamamoto, 1969) and the spore
vesicles (SV) (Gregg and Badman, 1970). At this time the α-prestalk
cells also differentiate, although they do not possess unique structures
by which they may be easily identified.

The presence of the PV and SV provided markers which were essential
to Gregg and Badman (1970) and Gregg (1971) in studies concerned
with cell differentiation and redifferentiation. Gregg (1971) dispersed
vegetative or early aggregate myxamoebae with a hair loop on agar.
These individuals dispersed as individual cells remained in isolation
for 24 hr but failed to synthesize the PV or SV markers. In a similar
experiment prestalk cells and prespore cells isolated from a slug were
maintained as individual cells for a period of 5 hr (Fig. 16). In this instance, the prestalk cells did not synthesize the prespore markers nor
did the prespore cells lose their PV's although SV's appeared to be lost.
Thus isolated cells do not have the capacity to synthesize or lose their
entire complement of characteristic organelles, indicating that they are
unable to undergo the normal transition or redifferentiation unless they
are in close association with other cells. Sakai and Takeuchi (1971) conducted a study of the fate of the prespore vacuole (PSV) in *D.
discoideum* prespore cells isolated from slugs by a Pronase–dimercapto-
propanol solution. They observed that 5 hr of isolation resulted in the
degradation of approximately 90% of the PSV in the prespore cells. The
loss of PSV was also correlated with the disappearance of a prespore
specific antigen detectable by fluorescent antiserum (Takeuchi and Sakai,
1971; Ikeda and Takeuchi, 1971). This process presumably is not a redifferentiation into prestalk cells but probably a dedifferentiation to the
vegetative stage.

Prestalk and prespore cell *groups* isolated from migrating pseudoplas-

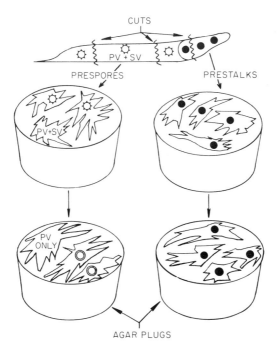

FIG. 16. Diagram indicating the failure of single isolated *D. discoideum* prestalk or prespore cells to redifferentiate into the opposite cell type. The inability of the cells to redifferentiate is revealed by the stability of the prespore vacuoles (PV) although spore vesicles (SV) disappear. The prestalk cells fail to synthesize either of these prespore components.

modia redifferentiate the appropriate missing cell type within a few hours. Thus the prestalk *group* synthesizes PV and SV in a certain proportion of the cells, while the prespore *group* loses PV and SV in the appropriate numbers of cells (Fig. 17). This regulatory process, of course, serves to ensure the construction of normal, proportioned fruiting bodies.

Obviously the prestalk and prespore cells establish an equilibrium in the intact slug which governs their proportions. This equilibrium is destroyed when the two cell types are separated and redifferentiation ensues. Consequently, it was anticipated that when prestalks or prespores were isolated as single cells they would respond by attempting to redifferentiate into the opposite cell type. The reduction of this theoretical possibility to practice, however, would only serve about the same degree of purpose found in a marooned sailor capable of metamorphosing into the object of his frustration.

Experiments in this laboratory indicate that prestalk or prespore cells isolated for more than 3 hr fail to form fruiting bodies following mechani-

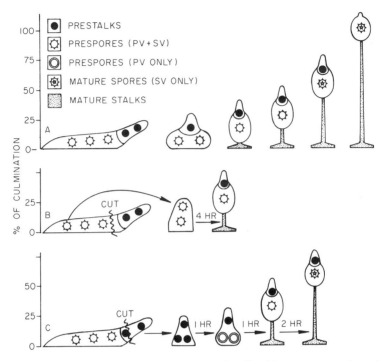

Fig. 17. Diagram of the differentiation of *D. discoideum* myxamoebae during normal development and of the redifferentiation of prestalk and prespore isolates. A. Differentiation of the cells under normal conditions. B. Prespore isolate indicating redifferentiation of a certain proportion of prestalk cells as indicated by the absence of prespore vacuoles (PV) and spore vesicles (SV). C. Prestalk isolates redifferentiating a certain proportion of prespore cells as revealed by the initial resynthesis of PV and subsequently SV.

cal recombination (Gregg, 1971). Thus it appears that the loss of critical substances is irreversible at this stage. Bonner *et al.* (1969) have determined that acrasin production in *D. discoideum* slugs falls to a lower level relative to the aggregation stage. Consequently, the failure of single cell isolates to redifferentiate may result from the loss of cell entities and intercellular communication which are normally maintained in multicellular situations. A marooned prespore cell, however, is not doomed to fungal celibacy if bacteria are present and cell division occurs. Prespore cells isolated under these circumstances lose their PV and SV within 8 hr and are considered to have returned to the vegetative stage (Gregg, 1971).

Loomis (1970) has composed a graph illustrating the pattern of accumulation of eleven enzymes during the development of *D. discoideum*.

The synthesis of ten of the eleven enzymes appears to depend on cell interactions following aggregation. The remaining enzyme, D-mannosidase, is synthesized continually from preaggregation through the culmination stages. If slugs are dissociated to single cells, D-mannosidase continues to accumulate emphasizing that certain syntheses do not require cell association. M. Sussman (personal communication, 1970) has observed that the enzymes UDP-galactose transferase, UDP-glucose pyrophosphorylase, UDP-galactose epimerase, and trehalose-6-phosphate synthetase in disaggregated *D. discoideum* migrating slugs remain at the same levels existing in the intact slug. With the exception of epimerase, following reaggregation additional amounts of the three enzymes are synthesized and accumulated. Epimerase, however, disappears during the reaggregation process. Subsequently, epimerase is resynthesized during the morphogenetic period in which it normally appears. It is interesting to note that both synthesis and degradation of certain enzymes depends on the reassociation of the cells.

The necessity of cell association in differentiation infers an interaction of the surfaces of the apposing cells. A freeze-etch study of the plasma membranes of *D. discoideum* was conducted by Aldrich and Gregg (1973) which described the changes in the diameter and frequency distribution of the plasma membrane particles. Vegetative myxamoebae membranes contain particles which have a mean diameter of 60 Å (Fig. 18A). During the subsequent morphogenetic stages the particles attain 110 Å in mean

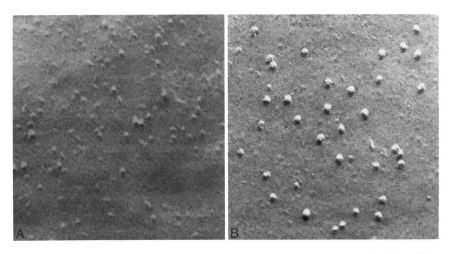

FIG. 18. Freeze-etch preparations of plasma membranes of *Dictyostelium discoideum* cells from the vegetative and preculmination stages. A. Vegetative myxamoebae. Mean particle diameter 60 Å. ×159,000. B. Preculmination stage unspecified cell type. Mean particle diameter 110 Å. ×159,000.

diameter (Fig. 18B). The decline in frequency of the small particles during morphogenesis suggest that they are combining to form the large particles. Since the plasma membrane is only 67 Å thick, the large particles
may project beyond the phospholipid bilayer. The particles are considered to be globular proteins which may be involved in cell adhesion.
Cyclic AMP which is known to increase adhesiveness between the cells
(Konijn et al., 1968) also induces the formation of large particles in the
plasma membranes of vegetative myxamoebae (Gregg and Nesom, 1973).
Singer and Nicolson (1972) proposed that unit membrane globular proteins respond to physical or biochemical stimuli by changing configuration which results in the transmission of effects to the cytoplasm.
Consequently, in *Dictyostelium* the globular proteins may compose the
cell contact sites and simultaneously induce the events which result in cell
differentiation.

B. Acrasin and Cell Differentiation

Cell differentiation, however, may be induced experimentally in isolated cells. Bonner (1970) has recently demonstrated that *D. discoideum*
vegetative myxamoebae would differentiate into mature stalk cells on
agar containing cyclic AMP. If small drops of myxamoebae are placed
on the agar, the cells will begin to migrate outward eventually forming
an outer ring and an inner ring. This movement occurs in response to
the phosphodiesterase activity of the cells which destroys the cyclic AMP
in the vicinity. Thus the cells are continually attracted to the higher
concentration of cyclic AMP at the periphery of the drop (Bonner *et
al.*, 1969). Approximately 24–48 hr under these conditions are required
for the middle ring of cells to differentiate into mature stalk cells.

Let us consider the factors associated with mature stalk cell differentiation under normal circumstances. A migrating pseudoplasmodium is characterized by anterior prestalk cells having high acrasin production
relative to the prespore cells. Slugs prior to migration do not exhibit this
sharp differential in acrasin synthesis (Bonner, 1949b). This finding suggests that either active acrasin secretion occurs following prestalk cell
differentiation or phosphodiesterase activity decreases in this region. The
prestalk cells subsequently require about the same period of time to
differentiate into mature stalk cells as those induced by cyclic AMP.
Under both circumstances the common factor is cyclic AMP in relatively
high quantities, although its effect is probably secondary. Otherwise, it
might be expected that cells on cyclic AMP agar would differentiate
within a shorter time period.

What is the effect of cyclic AMP on isolated myxamoebae which induces them to differentiate into mature stalks? This question may be answered only at a superficial level at this point. Under normal circumstances, a cell beginning to secrete acrasin will initiate a response from the surrounding cells which stream toward the point source of acrasin and slug formation normally ensues. Myxamoebae on cyclic AMP agar, however, lack a point source toward which aggregation is focused and consequently fail to accomplish cell association. Instead, the cells actively move outward, constantly increasing the intercellular distance until stalk cell differentiation occurs. Although the cells fail to aggregate, plasma membrane ultrastructural changes, perhaps necessary in cell differentiation, occur in isolated cells (Gregg and Nesom, 1973).

In theory, a single cell on plain agar may begin to secrete acrasin but in the absence of other cells this activity may not be sustained. Shaffer (1961) and Gerisch (1966) suggest that in *Polysphondylium violaceum* and *D. minutum*, respectively, founder cells function only for a limited time and then return to potential responders provided they do not succeed in initiating an aggregate. Therefore, independent cells on plain agar probably do not liberate acrasin in the quantities necessary to duplicate the environment offered by cyclic AMP agar and consequently cannot differentiate into stalk cells.

Prestalk cells are characterized by their relatively great expenditure of substrates such as protein (Gregg *et al.*, 1954). It is known that cyclic AMP acts as a messenger in a variety of systems in which the activation of metabolic enzymes is involved (Rasmussen, 1970). It is suggested that cyclic AMP may play a role in promoting mature stalk differentiation by inducing the necessary metabolic events in prestalk cells under either normal or experimental conditions. It might be postulated that prespore cells are prevented from differentiating into prestalk cells by maintaining a relatively high phosphodiesterase activity. Other than some preliminary experiments conducted in our laboratory, no information is available on the differential activity of the enzyme in the slug. These experiments were based on the degree of spreading by vegetative myxamoebae, prestalk, and prespore disaggregates on cyclic AMP agar. The observations revealed that vegetative myxamoebae > prespores > prestalks with respect to their spreading tendencies. This suggests, but by no means proves, the phosphodiesterase activity has decreased by the slug stage and that prespores have relatively greater activity than prestalk cells. An inhibitor of cyclic AMP-phosphodiesterase which appears by the aggregation stage in *D. discoideum* (Riedel and Gerisch, 1972) may account for the decline of activity in the slug.

C. Morphogenesis and Cell Differentiation

Although isolated mature stalk cell differentiation has been experimentally induced (Bonner, 1970), stalk cell formation under normal circumstances appears to depend on morphogenetic movements. Gregg and Badman (1970) determined that prestalk isolates which were disrupted periodically with a glass needle were capable of resynthesizing PV and SV in a certain proportion of the cells (Fig. 19). Although redifferentiation of prestalk cells into prespore cells occurred, mature stalk cells did not appear in the isolates. The prevention of normal morphogenetic movements and spatial relationships of the cells apparently resulted in this failure.

Spatial orientation does not appear to be necessary in certain instances as highly aberrant fruiting bodies composed of differentiated stalk and

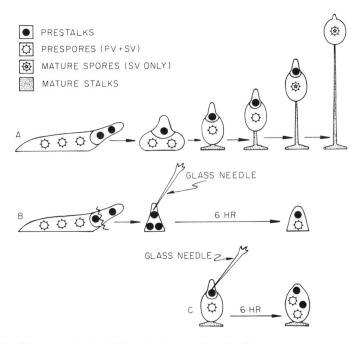

FIG. 19. Diagram of the differentiation of *D. discoideum* myxamoebae under normal and experimental conditions. Derived from Gregg and Badman (1970). A. Differentiation of the cells into mature stalks and spores under normal conditions. B. A prestalk cell isolate arrested mechanically for 6 hr revealing the redifferentiation of a proportion of the cells as indicated by the presence of prespore vacuoles (PV) and spore vesicles (SV). Mature stalk cell differentiation did not occur. C. An early culminate arrested for 6 hr revealing the failure of mature spore differentiation, which normally would have occurred during a similar period of time.

spore cells have been produced in *Dictyostelium* with EDTA (ethylene-diaminetetraacetic acid) (Gerisch, 1961). The *Dictyostelium* mutant, Fr-17, also forms an amorphous fruiting body containing mature stalks and spores (Sonneborn *et al.*, 1963). However, even in these instances, spore differentiation did not occur in the absence of stalk differentiation which supports the hypothesis that cell interaction is necessary in the culmination process.

Mature spore differentiation normally occurs at about the midpoint of the culmination process. By this time approximately 50% of the prestalk cells have differentiated into mature stalk cells. If culmination is prevented mechanically prior to the midpoint of culmination mature spore differentiation does not occur (Gregg and Badman, 1970) (Fig. 19). This suggests that mature spore differentiation is linked to stalk formation in some obscure way in *D. discoideum*. There is increasing evidence that synthetic activities within the cells and the particular morphogenetic state of a pseudoplasmodium are interrelated. Newell and Sussman (1970) have correlated the appearance and disappearance of several enzymes at particular morphogenetic stages. For example, during slug formation UDP-glucose pyrophosphorylase accumulates slowly and stabilizes at a certain level. UDP-galactose epimerase is not synthesized during this period. If the slugs are induced to begin culmination an additional synthesis of pyrophosphorylase occurs and epimerase begins to appear. Thus, the synthesis of certain enzymes appears to be linked with specific acts of morphogenesis. Newell and Sussman (1970) conclude that development in the slime molds depends on the ability of these cells to activate particular genetic switches when the necessity arises.

VI. Summary

A mature slime mold spore, upon germinating, yields a single myxamoeba whose subsequent progeny are capable of producing normal fruiting bodies. Obviously the developmental pathways of the homogeneous vegetative myxamoebae must diverge if fruiting bodies composed of two cell types are to be constructed. The α-prespore and α-prestalk cells in *D. discoideum* which eventually construct the mature fruiting bodies originate during the preaggregation stage (Takeuchi, 1969). Their appearance implies that the expression of a single genotype is regulated by means that are obscure at present. However, if the assumption is made that the myxamoebae develop asynchronously with respect to utilization or synthesis of various substrates, the basis for variability is established (Francis, 1969). It has been determined that all *P. pallidum* myxamoebae in liquid culture first differentiate into responder cells. Subsequently, acrasin-secreting cells begin to appear among the responders which may result from the asynchronous development of certain cells (Francis, 1965). Thus, cells arriving at a particular physiological stage secrete acrasin (α-prestalks?) and effect an aggregative response from the adjacent cells (α-prespores?). Consequently the variability that is initiated among the preaggregation myxamoebae results in cell association which is essential in the transition to a mature fruiting body. The α-prespore and α-prestalk cells complete differentiation into prespore and prestalks only following aggregation and pseudoplasmodium formation (Gregg and Badman, 1970). Neither cell type can differentiate among preaggregating cells maintained in isolation (Gregg, 1971).

Ultrastructural changes induced by cyclic AMP in the plasma membrane surfaces of *D. discoideum* occur during development (Gregg and Nesom, 1973). The appearance of large particles, presumably globular proteins, accompany the transition of the vegetative myxamoebae to the morphogenetic stages (Aldrich and Gregg, 1973). These proteins may be concerned with cell adhesion and the transmission of effects to the cytoplasm resulting in cell differentiation (Singer and Nicolson, 1972). The ability of isolated prestalk or prespore *groups* to redifferentiate the missing cell type further emphasizes the necessity of cell contacts and interaction of the unit membranes. It is conceivable that cell differentiation also depends on the mutual production and accumulation of specific materials. A pseudoplasmodium surrounded by a slime sheath may be capable of conserving critical amounts of the materials secreted within the cell mass. Perhaps the failure of isolated single cells to differentiate may be attributed to the absence of apposing unit membranes and the dispersal of their secretions throughout the substrate. Although the nature of these specific materials are obscure, one naturally occurring substance in cellular slime molds, cyclic AMP, has effects on cell differentiation which justify discussion. Cyclic AMP is associated with the appearance of adhesiveness in aggregating myxamoebae (Konijn et al., 1968) which may reflect the unit membrane changes observed. Cyclic AMP in agar has the potential of effecting the differentiation of individual myxamoebae into mature stalk cells (Bonner, 1970), although mature spores never appear under these circumstances. Mature spore differentiation, at least in wild-type *D. discoideum*, cannot ensue without mature stalk formation during the culmination process (Gregg and Badman, 1970). This further emphasizes that future progress in determining the mechanisms involved in cell differentiation in the cellular slime molds may depend on studies of their cell interactions.

Acknowledgments

The authors are indebted to Dr. Ilse Ortabasi and Miss Pamela High for their conscientious research assistance and meaningful discussions which contributed to the preparation of this chapter. In addition, acknowledgments are due Dr. John T. Bonner, Princeton University, for his valued criticism and advice relative to this manuscript.

References

Aldrich, H. C., and Gregg, J. H. (1973). *Exp. Cell Res.* (in press).
Beug, H., Gerisch, G., Kempff, S., Riedel, V., and Cremer, G. (1970). *Exp. Cell Res.* **63**, 147–158.
Bonner, J. T. (1947). *J. Exp. Zool.* **106**, 1–26.
Bonner, J. T. (1949a). *Sci. Amer.* **180**, 44–47.
Bonner, J. T. (1949b). *J. Exp. Zool.* **110**, 259–271.
Bonner, J. T. (1950). *Biol. Bull.* **99**, 143–151.
Bonner, J. T. (1957). *Quart. Rev. Biol.* **32**, 232–246.
Bonner, J. T. (1959). *Sci. Amer.* **201**, 152–162.
Bonner, J. T. (1970). *Proc. Nat. Acad. Sci. U.S.* **65**, 110–113.
Bonner, J. T. (1971). *Annu. Rev. Microbiol.* **25**, 75–92.
Bonner, J. T., Chiquoine, A. D., and Kolderie, M. Q. (1955). *J. Exp. Zool.* **130**, 133–158.
Bonner, J. T., Barkley, D. S., Hall, E. M., Konijn, T. M., Mason, J. W., O'Keefe, G., III, and Wolfe, P. B. (1969). *Develop. Biol.* **20**, 72–87.
Farnsworth, P. A., and Wolpert, L. (1971). *Nature (London)* **231**, 329–330.
Francis, D. (1965). *Develop. Biol.* **12**, 329–346.
Francis, D. (1969). *Quart. Rev. Biol.* **44**, 277–290.
Francis, D., and O'Day, D. H. (1971). *J. Exp. Zool.* **176**, 265–272.
Gerisch, G. (1961). *Arch. Entwicklungsmech. Organismen* **153**, 158–167.
Gerisch, G. (1966). *Arch. Entwicklungsmech. Organismen* **157**, 174–189.
Gerisch, G. (1969). *Eur. J. Biochem.* **9**, 229–236.
Gregg, J. H. (1956). *J. Gen. Physiol.* **39**, 813–820.
Gregg, J. H. (1965). *Develop. Biol.* **12**, 377–393.
Gregg, J. H. (1966). *In* "The Fungi" (G. C. Ainsworth and A. S. Sussman, eds.), Vol. 2, pp. 235–281. Academic Press, New York.
Gregg, J. H. (1971). *Develop. Biol.* **26**, 478–485.
Gregg, J. H., and Badman, W. S. (1970). *Develop. Biol.* **22**, 96–111.
Gregg, J. H., and Bronsweig, R. D. (1956). *J. Cell. Comp. Physiol.* **47**, 483–487.
Gregg, J. H., and Nesom, M. G. (1973). *Proc. Nat. Acad. Sci. U.S.* (in press).
Gregg, J. H., Hackney, A. L., and Krivanek, J. O. (1954). *Biol. Bull.* **107**, 226–235.
Hohl, H. R., and Hamamoto, S. T. (1969). *J. Ultrastruct. Res.* **26**, 442–453.
Ikeda, T., and Takeuchi, I. (1971). *Develop., Growth Differ.* **13**, 221–229.

Konijn, T. M., Van de Meene, J. G. C., Bonner, J. T., and Barkley, D. S. (1967). *Proc. Nat. Acad. Sci. U.S.* **58,** 1152–1154.

Konijn, T M., Barkley, D. S., Chang, Y. Y., and Bonner, J. T. (1968). *Amer. Natur.* **102,** 225–233.

Krivanek, J. O., and Krivanek, R. C. (1958). *J. Exp. Zool.* **137,** 89–115.

Loomis, W. F., Jr. (1970). *J. Bacteriol.* **103,** 375–381.

Maeda, Y., and Takeuchi, I. (1969). *Develop., Growth Differ.* **11,** 232–245.

Miller, Z. I., Quance, J., and Ashworth, J. M. (1969). *Biochem. J.* **114,** 815–818.

Newell, P. C., and Sussman, M. (1970). *J. Mol. Biol.* **49,** 627–637.

Raper, K. B. (1940). *J. Elisha Mitchell Sci. Soc.* **56,** 241–282.

Rasmussen, H. (1970). *Science* **170,** 404–412.

Riedel, V., and Gerisch, G. (1971). *Biochem. Biophys. Res. Commun.* **46,** 279–287.

Sakai, Y., and Takeuchi, I. (1971). *Develop., Growth Differ.* **13,** 231–240.

Shaffer, B. M. (1958). *Quart. J. Microsc. Sci.* [N.S.] **99,** 103–121.

Shaffer, B. M. (1961). *J. Exp. Biol.* **38,** 833–849.

Singer, S. J., and Nicolson, G. L. (1972). *Science* **175,** 720–731.

Sonneborn, D. R., White, G. J., and Sussman, M. (1963). *Develop. Biol.* **7,** 79–93.

Sonneborn, D. R., Sussman, M., and Levine, L. (1964). *J. Bacteriol.* **87,** 1321–1329.

Sussman, M., and Lee, F. (1955). *Proc. Nat. Acad. Sci. U.S.* **41,** 70–78.

Takeuchi, I. (1963). *Develop. Biol.* **8,** 1–26.

Takeuchi, I. (1969). *In* "Nucleic Acid Metabolism Cell Differentiation and Cancer Growth" (E. V. Cowdry and S. Seno, eds.), pp. 297–304. Pergamon, Oxford.

Takeuchi, I., and Sakai, Y. (1971). *Develop., Growth Differ.* **13,** 201–210.

5

Metabolism, Cell Walls, and Morphogenesis

STUART BRODY

I. General Morphogenesis

A. Introduction

Many authors have divided the field of developmental biology into the study of differentiation and the study of morphogenesis. Although certain areas of investigation are difficult to categorize into either group, a useful distinction between these two, similar to that suggested by Davis (1964), is to equate differentiation with selective gene action and morphogenesis with the actual shaping of three-dimensional structures from the gene products. In this chapter I will discuss the relation of genes and gene products to morphogenesis but will not focus on the translational or transcriptional controls of gene activity. In this regard, some of the questions dealing with the interrelationships of genes, enzymes, structures, and phenotypes are listed below. Although this is not a complete list, it is hoped that these questions and a comparison of approaches which can be employed to answer them might be more informative and stimulating than a general review of the literature. The answers to questions about the chemical basis of morphogenesis in most systems have either been scarce, not described in sufficient detail, or beg the question altogether. In a few instances, detailed answers to these questions already exist, such as in the study of viral morphogenesis. This subject has been well reviewed by Levine (1969) and Wood *et al.* (1968) and included in certain excellent overall reviews, such as "Self-Assembly of Biological Structures" (Kushner, 1969), and therefore will not be reviewed here. In other studies, such as those concerned with microbial cell walls, the answers are interesting, but not as detailed as in the viral systems. However, due to my experiences, this article will emphasize what is known about fungal cell walls and morphogenesis from a biochemical genetic standpoint.

B. Questions about Shape Determination

1. *Do Enzymes Play a Major or Minor Role in the Formation of a Given Structure?*

It is already known that aspects of certain processes or even entire morphogenetic sequences, such as tobacco mosaic virus formation (Lauffer and Stevens, 1968), are more critically dependent on physical-chemical interactions than on enzymatic actions at the site of structure formation. Perhaps a more meaningful way to phrase this question would be to ask what is the relative importance of both physical-chemical ordering and catalytic activities? The class of physical–chemical interactions can be subdivided into those molecules that spontaneously assemble themselves (self-assembly) vs. those that require additional surfaces for assembly into structures. This latter group would need the transient presence of morphopoietic factors (Laemmli *et al.*, 1970; Yanagida *et al.*, 1970), either in the form of nonenzymatic proteins or other macromolecules, which would not themselves be incorporated into the final structure. The class of catalytic activities can be subdivided into those reactions which occur during or after the formation of the initial structure. An example of this latter type would be a hydrolytic cleavage at the periphery of newly formed structures to remove terminal sections of exposed polymers, a sort of trimming. In considering the role of enzymes in shape determination, one should also be aware of situations where enzymes are necessary but not sufficient for structure formation. For example, in blood clotting the formation of the clot is directly under enzymatic control, but the site of the fibrin clot is not initially determined by the enzymes themselves. Therefore, it is clear that this whole first question is complex and leads to a series of other questions, each requiring illustration and clarification beyond the scope of these introductory paragraphs.

2. Can Shape-Determining Processes Be Completely Described by a Series of Enzymatic and Nonenzymatic Assembly Steps?

In other words, a complete chemical description of the enzymes and polymers that participate in each step might not include a description of those physical forces, such as turgor pressure, which can influence the site or shape of a structure. In some instances, certain mechanical forces acting on an existing structure could be the key-determining factors of the observed patterns.

3. Is There Strict Genetic Control of the Complete Morphology of All Cellular Structures?

We usually take it for granted that each aspect of the gross morphology of a structure is genetically determined; however, there may be a certain amount of nonuniformity at a lower level in terms of the individual components or molecules in a structure. One can imagine that a particular amount of variation in membrane structure may exist, for example, due to local ionic effects and other environmental influences. In fact, Kellenberger (1968) has postulated that local nongenetic effects may even alter *gross* phenotypes, such as the size of a virus particle. He has called these altered viruses "physical variants" as opposed to informational (mutant) variants. Therefore, it would be interesting to know, for any given structure, the relative importance of direct genetic control vs. local environmental influence.

4. What Are the Primary Cellular Activities of Genes Directly Involved in Shape-Determining Processes?

This question refers to those enzymes or proteins which actually form structural precursors such as polymers, but not to those enzymes involved in the production of ATP or other intermediates. One would also like to know if the particular gene products concerned mainly with shape determination have any secondary functions, such as the transport of materials in or out of the cell.

5. *Are Integrated Morphogenetic Processes the Expression of Corresponding Integrated Patterns of Enzyme Synthesis, or Are They the Expressions of Interlocking Regulation of the Activities of Existing Enzymes?*

The integration of genic activity is often thought of in terms of a "chromosome clock" containing repressors with different stabilities which activate certain genes or groups of genes independent of any input signal. This type of regulation could be either the simple cascade type of regulation or some more complex system of related effectors. On the other hand, the control of morphogenesis by a feedback type of regulation assumes that the bulk of the enzymes necessary for the formation of a structure are always present in excess amounts and that the activity of these enzymes is then controlled at the pertinent site and time by small molecule allosteric regulators. It would be interesting to know which morphogenetic processes were controlled by one of these two mechanisms, or by a combination of these two types of regulation.

All the above questions are directed in one way or another to the relationship between chemical events and shape determination. The reciprocal relationship, how shape influences metabolism, may be even more complex. For example, one can imagine that increased concentration of metabolites occur inside certain areas when cells are arranged as cylinders or balls as compared to the situation where these excretory products would diffuse away from a simple sheet of cells. This subject will not be considered in this review, even though it may be of considerable importance in morphogenesis and development.

II. Possible Approaches to Answers

A. GUIDELINES FOR THE SELECTION OF AN ORGANISM

For the study of morphogenesis, the most important quality that an organism should possess is a well-defined structure or structures which undergo visible shape changes during growth and development. The size of these structures need not be a factor in selecting a particular organism, since potentially interesting structures can be found over a large size range from the entire exterior surface of an organism down to a small, specialized cellular organelle. In a similar way, the size of an organism should not be an overriding factor, within certain obvious limits. However, the simplicity of a structure, in terms of its visual description and number of major components, would be a reasonable operational consideration. This would also apply to structures that were basically two-dimensional vs. three-dimensional. Perhaps the most important consideration would be the many types of experimental manipulations that can be performed on an organism, and these are discussed below.

Primary consideration should be given to the qualities of an organism that allow particular *genetic* experiments to be performed. For example, mutant types can be obtained from many organisms, but it is clearly beneficial to work with an organism in which *selective* techniques can be applied toward increasing the number and variety of mutants. Second, it is somewhat easier to obtain mutants in haploid organisms than diploids since recessive mutations are immediately expressed in haploids. However, there may be certain subtle advantages to working with diploids, such as the ability to easily study gene dosage effects. If one already has a collection of mutants, it may be helpful if linkage relationships can be ascertained, even though the resulting map does not necessarily lead to any further understanding of the primary effect of any of the mutations. In addition, a given mutant can give rise to revertant and suppressor strains provided that large numbers of the organism can be easily and routinely handled. In fact, the ability to obtain rare genetic modifiers, such as conditional revertants, may be the most important tool for studying pleiotropic gene effects on morphogenesis (see end of Section IV). Other genetic capabilities, such as fine structure mapping or the determination of dominance–recessive relationships, are helpful but probably optional. Another optional, but practical consideration is the ability to preserve hundreds of mutant strains easily as stab cultures or lyophilized cultures. In summary, even though the full range of genetic manipulations may not be employed on a given organism, the potentially powerful applications of this approach to the study of morphogenesis should be kept in mind.

The focus in *biochemical* studies is primarily quantitative. Enough biological material must be comfortably and repeatedly obtained to allow, for example, the appropriate characterization of a given enzyme. The amount required will vary, depending on the nature of the investigation. If studies on peptide patterns (fingerprinting technique) are to be performed, then at least 100 gm dry weight are needed initially. This estimate is derived from the assumption that soluble protein constitutes approximately 30% of the dry weight of an organism, that a given enzyme will constitute approximately 0.1% of the total protein in a crude extract, that 20% overall recovery of the total enzymatic activity is possible, and that at least approximately 5 mg of protein are needed to do initial studies on peptide patterns. This minimum can be reduced when microtechniques are available for studies of peptides and subsequent sequence analyses, or when conditions can be found to raise the level (derepression) of the particular enzyme in question. This minimum can be reduced if the level of analysis does not require studies on the primary structure of a protein, but will be satisfied with electrophoretic or immunochemical analysis on purified preparations. If the analysis only requires measurements of enzyme activity in crude extracts, then even less material is required (possibly < 50 mg). If the investigator does not measure enzymes at all, but is more concerned with assaying the steady state levels of the low molecular weight intermediates, then the minimum estimate of dry weight required would be of the order 0.1–1.0 gm. Clearly, the use of certain microtechniques and isotopic labeling studies can lower this estimate also. If the investigator is studying the concentration of a morphogen by *in vivo* assays, then one should be concerned with the accuracy and the time needed to perform the assay as well as the ability to gather enough morphogen to perform a standard biochemical analysis of this material. It may seem trivial to emphasize quantity so strongly, but one often sees interesting investigations that cannot be further analyzed at molecular levels for this reason alone. Certain analyses might require, for example, a football field of a monolayer of cells or thousands of man-hours to dissect out the required amounts of a specialized bit of tissue. Therefore, the quantitative aspects should not be overlooked when considering a given organism, even though the type of analyses described above may require considerable foresight.

A third way of judging the qualities of an organism would be from the *physiological* standpoint. In this respect, it is clearly beneficial to study an organism which can be manipulated in many ways, such as transplantation experiments, tissue reaggregation experiments, etc. It is also very helpful to be able to employ a range of growth temperatures and other environmental controls and, if possible, a defined media. These general physiological qualities of an organism appear to be helpful to an investigator, whereas the important anatomical considerations that an organism should possess are not as easily defined. For example, it is not always obvious why the organism of choice should be a eukaryote with well-defined membranes, chromosomes, and mitochondria. An investigator must strike a balance between the attractive features and relative simplicity of the lower organisms (prokaryotes) vs. the stronger similarities among the higher organisms (eukaryotes). Certain features of eukaryotes are useful, such as the ability to describe the chromosome topography in relation to studies of gene regulation. However, this feature does not appear to be as critical to the studies of the role of genes in morphogenesis. Other qualities that may also be considered optional would be the ability to obtain synchronous cell division and the appearance of distinct tissue types in an organism. For certain other investigations, these qualities would, of course, be indispensable. Finally, the type of life cycle of an organism would not appear to be an overriding factor, although the duration of that cycle might easily affect the patience of the investigator. In summary, one would like to be able to investigate an organism with all of the above properties which allowed every type of conceivable experimental manipulation to be performed. Since such a perfect laboratory organism probably does not exist, it would appear that some choices and compromises must be made. The preceding points can only serve as a bare framework for such choices, and their order of presentation—genetic, biochemical, and physiological—reflects only my experience and biases.

B. METHODOLOGY OF APPROACHES

1. *The Genetic Approach*

The purely genetic approach to the problems of morphogenesis has primarily involved the selection and/or the isolation of mutants. These mutants can be arbitrarily divided into at least three classes. The first class contains the lethal mutants, such as those of *Drosophila* studied by Hadorn (1961) and those of other organisms as studied by Gluecksohn-Waelsch (1963), which have been used to determine the time of action of particular genes. A second class contains the conditional mutants such as those first studied by Horowitz and Shen (1952) in *Neurospora* and now includes many temperature-sensitive phenotypes of *Drosophila* obtained by Suzuki (1970), the temperature-sensitive mutants of yeast isolated by Hartwell (1967), the temperature-sensitive mutants of *Dictyostelium discoideum* isolated by Loomis (1969), and those in bacteriophage T4 isolated by Edgar and Lielausis (1964). A third class includes those viable morphological mutants produced by point mutations, or small intragenic aberrations, with altered phenotypes which are relatively insensitive to environmental conditions. If these different types of mutants are collected and analyzed genetically, then a great deal of information can be obtained. We can tell about the clustering of mutants (from mapping data), about their complementation properties, about dominance–recessive relationships, and about gene interactions. In addition, if the analysis is carried into the physiological realm, one can often determine roughly the time of action of a given gene by varying the time of temperature shift for a mutant strain (Foster and Suzuki, 1970). This technique enables one to determine approximately when a gene product is produced or when it is needed in the life cycle of an organism (Hartwell *et al.*, 1970). This same type of penetrating analysis can also be obtained by using the experimental techniques which enhance mitotic recombination (Stern, 1969). However, usually none of these genetic or environmental manipulations provide significant clues as to the primary gene products of these mutant alleles. So, in this sense, a purely genetic analysis is limited in its scope.

On the other hand, if one could select for specific mutants blocked in a given developmental pathway, as Fristrom (1970) pointed out, then he could concentrate on specific pathways or facets of morphogenesis. This type of intensive analysis would probably simplify the eventual biochemical studies of these mutants, since a clue to the defect in one mutant should presumably lead to the elucidation of the defect in all of the others. In the absence of selected mutants one attempts to detect by biochemical screening programs the primary enzymatic deficiency that produced the randomly selected abnormal phenotype. This particular approach, analogous to the pediatrician's approach to inborn errors, will be discussed in more detail in Section II,C below. In any event, both genetic approaches, either the isolation of random mutants or the selection of specific mutants, have two powerful advantages. The first clear advantage is that the isolation of mutants blocked in a given step indicates that a particular *in vivo* change has a serious effect on an organism. It is never clear when a change is observed by an *in vitro* assay, whether this change has any significance to the physiology of the organism or whether the interlocking metabolism of the organism can easily compensate for this change. The second clear advantage is that the presence of mutant organisms gives the investigator an experimental handle for the natural alteration of the phenotype as contrasted with pharmacological alterations employing antibiotics, analogs, or chemical poisons. This advantage is quite noticeable when one contrasts the single primary effect of most mutations with the multiplicity of effects that are often produced by chemical perturbations.

2. The Biochemical Approach

The strict biochemical approach to the study of morphogenesis has been basically a descriptive one, differing only from a cytological approach in that the description has been on a molecular level and the vocabulary has been chemically defined. It is possible to gather enormous amounts of basic and helpful biochemical information about a particular structure and/or shape-determining process without really elucidating the molecular species that actually influence the emergent structure. In other words, the information obtained may describe the gross composition of a structure or may indicate the localization of an enzyme to this structure without adequately describing the physical-chemical forces that determine this structure. Perhaps, this type of information should be classified as correlative or descriptive information rather than causal or analytical information. For example, the description of pertinent proteins usually focuses on the role of the enzymatically active ones as opposed to structural proteins or peptides. This focus usually describes enzymes in terms of their specific activities, their stabilities, their localization within the cell or organism, their different electrophoretic forms (isozymes), their size, aggregation state, coenzyme content, kinetic properties, substrate specificity, inhibitor effects, etc. All these properties indicate a great deal about a particular isozyme, but may not indicate how this isozymic form is actually involved in the formation of a given structure or in a given differentiation of the organism. Similarly, nucleic acids are described in terms of nucleotide composition, size, density, location in the cell, time of appearance, hybridization properties, etc., all of which require considerable effort. These studies are important for their intrinsic value as well as for the basic background information they provide. Occasionally, there is even a remarkable payoff and one can relate genes, gene products, and phenotypes, as in the example of the ribosomal RNA deficiency in the bobbed mutants of *Drosophila* (Ritossa *et al.*, 1966). Hopefully, our eventual understanding of nucleic acid regulation, as indicated by initiation factors, psi factors, and many undiscovered factors, will give rise to a series of illuminating experiments pertinent to morphogenesis. The other biochemical components involved in structure formation, such as polysaccharides, lipids, and structural proteins, can also be exhaustively described. In some instances, the analyses of these molecules are more pertinent to morphogenesis, especially where these compounds are the principal components of a structure. But again, one is faced with the dilemma of deciding which components are critical and which are secondary. This question can often be answered by noticing the effects produced by enzymatic degradation or chemical removal of particular parts of a structure. If a component can be designated as critical in this way, then one can look at the time of appearance of this component, its mode of synthesis, as well as the regulation of the pertinent gene products. Another type of biochemical study involves the determination of the patterns of the low molecular weight intermediates. Even when a thorough analysis is performed as to the types, amounts, localization, flow, excretion, etc., it still may not be clear what role a given small molecule plays in the whole scheme of things. However, the small molecule patterns are very useful as preliminary clues (as indicated in Section II,C below) in many instances, even if not as an end unto themselves.

3. *The Cytological Approach*

The cytological approach describes a given organism in terms of its overall size and shape, as well as its tissue types, their relationship to each other, and their size and shape. This type of analysis has been extended to the cellular and subcellular levels, where the intracellular topography is described in terms of organelle distribution, membrane-bound structures, tubules, etc. The internal organization of cells and their visible surface properties have also been studied in a kinetic sense, via time-lapse photography. All the numerous techniques and painstaking cytological efforts are very necessary for the proper understanding of structure determination even though this approach, when employed alone, rarely yields biochemical clues as to which particular enzymes, substrates, or building blocks are involved in a given structure.

4. *The Physical Approach*

The physical approach has employed a variety of different experimental manipulations in order to determine their effects on the morphology of a given structure. These physical perturbations have included temperature changes, as well as changes in the surrounding environment, such as osmotic strength, pH, or the gaseous composition and/or pressure. Other parameters that have often been varied are the intensity, wavelength, and polarization of light, the strength of the surrounding electric field, and mechanical deformations. An important physical manipulation which proved useful is the transplantation of tissues, cells, or parts of cells to a different biological environment. This new environment may be either a different organism, a different tissue, a different cell line, an early stage embryo (Gurdon and Woodland, 1968), or just a reorientation of the existing material, such as an 180°-rotation of a tissue fragment. The use of microinjection techniques has also been employed (Wilson, 1961) for determining the role of various subcellular particulate fractions. All these techniques are helpful for the analysis of development in general, but again they rarely yield significant biochemical clues as to the nature of the critical molecules in the morphogenesis of a given structure.

C. Possible Methods for the Detection of the Primary Biochemical Change Due to Point Mutations

This section is included since morphological mutants of many organisms exist and these mutants usually contain structures altered in interesting ways. In order to employ these mutants usefully, ideally one should first know the primary deficiency in these variants and then proceed from that point in a rational manner. Since most of these morphological mutants were isolated from random mutant hunts or other screening techniques, but not via selective techniques, it is usually not clear what area of metabolism may be altered by the mutation. If one cannot visualize the mutationally altered component by microscopy, or if one is a little rusty at throwing darts at large wall charts of the metabolic pathways, then he must rely on some sort of a biochemical screening program.

There are at least six different basic approaches to the determination of the primary biochemical lesion. The first approach involves the analyses of nucleic acids by pulse labeling and/or hybridization techniques. Since present techniques probably would not detect single nucleotide changes, this approach would not be widely applicable. Hybridization techniques were successful, however, in elucidating the molecular basis of the bobbed mutants of *Drosophila*, since much more than a single nucleotide change was involved (Ritossa et al., 1966).

A second approach is to study *in vitro* protein synthesis with a reconstituted cell-free system. Here, one has a chance of detecting altered ribosomes, supernatant factors, tRNA, or amino acid activating enzymes. If one of these components is found to be altered, then well-known procedures can be used to track down the inadequacy to one molecular species. Elegant examples of this approach have been provided by Neidhardt (1966), as well as by Hartwell and McLaughlin (1968).

A third approach often employed is to assay enzyme levels in crude extracts. One is immediately confronted with the problems of how many enzymes should be measured, which enzymes should they be, and under what conditions they should be assayed. Even if the mutationally altered enzyme was fortuitously assayed, one would probably miss those mutants with enzymes of poorer substrate affinities (K_m type) since a substrate excess is part of the standard assay mixture. One might also overlook a "feedback negative" type of mutant if the proper low molecular weight allosteric effector was not present in the reaction mixture. Both of these objections can be partially overcome by a more detailed study of the kinetics of an enzyme. However, this study naturally requires more purified preparations and is then not a very rapid screening procedure. Perhaps the trump card for this approach is the use of electrophoresis for the detection of possible differences. In this technique, one may look at the patterns after electrophoresis on starch blocks, polyacrylamide gels, acetate strips, or via immunoelectrophoresis. Most of these techniques, now so widely used, rely on recognizable electrophoretic mobility changes. This requirement may exclude a significant fraction of altered enzymes, the exact fraction depending on the particular enzyme being studied. A second problem with this approach is that often one notices changes in the protein banding patterns with great enthusiasm, but then cannot identify the enzymatic activity associated with the altered protein bands. A third problem is the occurrence of dozens of differences in isozyme patterns in a mutant vs. the wild type (Bienengräber *et al.*, 1968). This, too, is frustrating, since it is not clear which enzyme has been altered as a direct result of the mutation and which enzymes are changed due to the pleiotropic effects of this mutation.

A fourth general approach is to scan mutants with respect to the composition of the major macromolecular components. This technique involves selective isotopic labeling followed by fractionation into the various classes of molecules, such as lipids, proteins, nucleic acids, and polysaccharides (Roberts *et al.*, 1955). This approach might give some clues as to the general nature of the mutation, but probably very few specific clues.

A fifth approach which has been successfully applied in several laboratories is to concentrate on the types and quantities of the low molecular weight intermediates. The rationale behind this technique is that these small molecules should reflect the state of the large molecules (enzymes and polymers). If the *in vivo* activity of an enzyme is considerably lower, one might expect *in vivo* substrate accummulation and/or lowered levels of the product of the enzymatic reaction. The actual mechanics of this approach are to extract as many of these molecules as possible and to separate them via a variety of chromatographic techniques. The proper identification of the spots or peaks requires colorimetric procedures, such as ninhydrin or $AgNO_3$ staining, UV absorption spectra, etc. If additional spots appear or disappear relative to the wild type, then these unknowns can be identified by the standard techniques of organic chemistry. This identification is obviously far easier than attempting to identify the enzyme activity of an unknown protein of altered mobility. If the small molecule pattern changes can be deciphered, one can use the enzymological approach with more acumen, since he then has significant clues as to the nature of the lesion. The drawbacks to this approach should also be pointed out. Some small molecule changes can go undetected if these molecules are not easily extracted, are not detected by the general staining procedures, or are present in very minute quantities. Second, substrates may not accumulate if they can be siphoned off by other cellular reactions, or if they do accumulate, they may merely reflect a pleiotropic effect and may not be closely related to the primary mutational alteration at all. Finally, this method only measures the total content of an intermediate and does not indicate the localization of these molecules, a critical fact for most cellular processes.

The sixth approach might be entitled the euphenic approach, and it involves the experimental improvement of an aberrant phenotype. This approach involves attempts to specifically reverse a mutant phenotype by the addition of one or a series of compounds. Although it is often a tedious and an inelegant operation, it can yield very direct clues as to the biochemical nature of a lesion. This method is obviously very similar to that employed by Beadle and Tatum (1941) for detecting unknown auxotrophic requirements. It should also be pointed out that the reciprocal approach, production of a mutant phenotype in a genetically normal organism (phenocopy),.has also occasionally been successful in yielding significant clues as to a biochemical step in morphogenesis (Rizki, 1961; Fristrom, 1965). In any event, the whole process of tracking down unknown enzymatic lesions is not quick, easy, or foolproof, and it is probably not for inexperienced biochemists.

III. The Cellular Morphogenesis of Certain Microorganisms

A. General Approaches

The cell walls of microorganisms carry out many functions other than serving as a protective barrier. They appear to be involved in stabilizing many organisms against dramatic osmotic changes in the environment (Livingston, 1969), in uptake of nutrients, in localizing extracellular hydrolytic enzymes (Chung and Trevithick, 1970; Trevithick and Metzenberg, 1966), and in the storage of reserve materials (Sussman et al., 1957). Some components of cell walls may also serve in the recognition processes of friendly microorganisms, such as the glucan-mediated mating reactions of yeast (Crandall and Brock, 1968). In this chapter microbial cell walls are singled out because of the large amount of information already known, the relative ease of manipulation of the microorganisms, the apparent basic role of the cell wall in shape determination, and my personal contact with this field.

Studies on the structure and function of microbial cell walls have usually focused on the role of the individual components in the cell wall. These components may be extracted from the intact organism or from purified cell wall material. Purified cell wall material has been used in order to reduce the cytoplasmic contamination as much as possible. The techniques used to accomplish this purification are varied and may consist of extensive washing of disrupted cell sediment with water or a concentrated salt solution or treatment of lyophilized material with detergent solutions. Unfortunately, the criteria that have been employed to determine the complete removal of cytoplasm often have been less than stringent and many cell wall preparations were probably impure. On the other hand, in those cases where the known cytoplasmic components were present only in trace amounts ($<0.1\%$), there is a possibility that some of the loosely bound and pertinent cell wall components may have been removed also. Therefore, investigators of purified cell walls may also be caught in the usual biochemical dilemma of analyzing chemically pure but biologically incomplete material.

Leaving aside this predicament, the first step in the characterization of cell wall components is the determination of the overall chemical composition of the entire cell wall. Once this analysis is complete, then the usual approach to structural studies involves the selective chemical or enzymatic removal of various components from the isolated cell wall. After removal and chemical characterization of some components, one naturally looks for differences in the visible appearance of the remaining cell wall and then tries to infer the role of the extracted component. Although this approach has been rather successful, the relative non-specificity of the chemical methods has often obscured the results. For instance, one needs to know exactly which of the many bonds that are broken actually led to the release of the component. Second, one needs to know if and how the remaining cell wall structure has been chemically modified. It is quite possible that when certain polymers are deacylated, or certain internal bonds are broken, significant changes are produced in the structure of this matrix even though these polymers remain in the cell wall. This approach can be extended by the use of enzymes, particularly highly purified enzymes of known specificity. The type of conclusions that might arise from the enzymatic or chemical approach would be that a given component of the cell wall must be removed before a second component can be solubilized. If this type of information can be gained, then an attempt can be made to determine which components are critical to the structure and what type of chemical interactions (covalent, electrostatic, etc.) actually keep the different components together.

A second general approach to cell wall structure studies has employed organisms with mutationally altered cell walls. This mutational approach has certain advantages in that it sidesteps the problem of steric hindrance—the inaccessibility to enzymes of certain bonds buried in the interior of the cell wall. Mutant cell walls may have new sites exposed to enzymatic attack and in this way allow the removal of interior components of the cell wall without strong degradative treatments. If certain cell wall components of mutants cannot be assembled properly into the existing matrix, then these components may be found in the media, often in a relatively undegraded form. This approach in combination with the straight chemical attack has often given information about the amount, type, localization, and role of an altered component. It has also indicated the type of effects produced on the cell wall due to the variation over a wide range (increased or decreased percentage) of a given component. In this respect, one must be careful to note those effects produced by compensatory changes such as the higher level of component number one due to a mutationally decreased level of component number two. One should not automatically attribute any observed changes in the cell wall properties to an increase in component number one. However, if many different mutants are available, then one can roughly deduce the variations of the amounts of components, or combinations of components, that are compatible with a viable organism.

A third approach has involved the analysis of the surface properties of the cell wall. This technique has employed specific antibodies, either ferretin labeled or conjugated with a fluorescent dye (Marchant and Smith, 1968) or certain newer techniques, such as specific phytohemagglutin reagents (concanavalin) coupled to dyes (Tkacz et al., 1971). The use of these reagents along with autoradiographic techniques and kinetic analyses allows one to gather information about whether a given chemical group is exposed on the surface of the cell wall, about the distribution of these antigens or polymers on the surface, and about their rate and order of appearance at the surface.

B. ASPECTS OF BACTERIAL CELL WALLS

Almost every approach and combination of these approaches have been employed for studying bacterial cell wall structure and synthesis. An excellent review by Osborn (1969) covers the chemical details of the surface layers of gram-negative and gram-positive bacteria. A more recent review by Ellar (1970) deals with the surface properties and shape-determining properties of bacterial cell walls, while a review by Rogers (1970) deals with the dynamics of bacterial growth and cell wall formation. It is interesting to note that there are two different modes of polymer synthesis in bacteria. One is the synthesis of polymers by the addition of monomers, either directly from a nucleotide precursor such as uridine diphosphoglucose or via a lipid-bound intermediate (mannosyl-1-phosphorylpolyisoprenol). The other is via the syntheses of small chains inside the cell such as the disaccharide pentapeptide nucleotide which is then added to the existing cell wall matrix. However, these two different modes can be viewed as similar if the complex nucleotide is considered to be a monomer in that it is also the repeating unit of a polymer. An interesting facet of this work is that the information needed for arranging the proper sequence of a complex monomer is not in a direct template of any kind but is due to the specificity of a battery of enzymes, each one adding a monomeric unit in a fixed sequence. In addition to the intracellular enzymes involved in bacterial cell wall biosynthesis, there may be a cell wall assembly enzyme actually localized in the cell wall (Ghuysen et al., 1968). This enzyme catalylzes a cross-linking transpeptidation reaction which is thought to link the complex monomers to the existing cell wall. It is speculated that the loss of a terminal amino acid residue during this reaction may provide the energy for this reaction on the outside of the cell. Further work on the role of enzymes in cell wall formation and cellular growth (Brown and Young, 1970; Ghuysen, 1968; Higgins and Shockman, 1970; Schwarz et al., 1969) might reveal more of the details of the assembly of the wall and of the intercalation of newly synthesized components.

Other aspects of bacterial morphogenesis that are due to cell wall changes or are suspected to be involved with cell wall alterations are the rod–sphere transformations (Krulwich and Ensign, 1968), the formation of chains, L cells, and minicells. In addition, the formation of a proper cell wall is probably very important for the functioning and synthesis of certain other surface organelles of bacteria, such as pili (Brinton, 1965), stalks (Schmidt and Stanier, 1966), and flagella (DePamphilis and Adler, 1971). Particular properties of cell walls have also been implicated in bacteriophage attachment (Lindberg and Hellerqvist, 1971) and in the uptake of transforming DNA (Tomasz, 1968) as well as in the formation of the septum during cell division (Hirota et al., 1968). Another extracellular bacterial polysaccharide which influences the visible morphology of the bacterial colony but is distinct from the cell wall is the slime layer of E. coli described by Markovitz (1964). He has reported the structure of this material, its mode and regulation of synthesis, and the effects of mutations on its biosynthesis (Lieberman and Markovitz, 1970; Lieberman et al., 1970).

Still another area of bacterial morphogenesis and development in which the external surface layers play an important part is sporulation. In the spores of bacilli, the presence of unusual compounds, such as dipicolinic acid, presumably associated with the cortical layers confers the heat-resistant properties to the spore (Halvorson and Swanson, 1969). Mutations which affect the synthesis of the cortex significantly alter the morphological and physiological properties of the spore (Pearce and

Fitz-James, 1971; Freese *et al.*, 1970). Numerous other mutations are known which may not directly affect cell wall formation but do affect sporulation and germination in these organisms (Balassa and Yamamoto, 1970; Doi *et al.*, 1970; Szulmajster *et al.*, 1970). Many of these changes have been reviewed recently (Freese *et al.*, 1968; Kornberg *et al.*, 1968). It is interesting to note that no qualitative differences have been found between spores and the mother cell with respect to DNA, RNA, numerous enzymes, and many small molecules (Nelson *et al.*, 1969; Hanson *et al.*, 1970). However, some significant differences have already been found for enzymes involved in cell wall formation (Tipper and Pratt, 1970), for RNA polymerase (Losick and Soneshein, 1969), and for other enzymes (see recent review by Hanson *et al.*, 1970). The study of sporulation encounters the same problem of physical localization (compartmentalization) of small and large molecules within one cell as does the study of the morphogenesis of most of the single-celled organisms. Until solutions to this problem can be devised, such as those suggested by Pitel and Gilvarg (1971), compartmentalization may very well be the most critical stumbling block toward a reasonable understanding of morphogenesis in these organisms.

C. Fungal Cell Walls

A direct relationship between cell walls and fungal morphogenesis has been suspected for a long time (Nickerson, 1964; Nickerson *et al.*, 1961) and thoroughly reviewed by Bartnicki-Garcia (1968) and Aronson (1965). Many different fungi have been studied, and in almost all of the cases where morphological changes have been described, significant changes in the composition of the cell wall were also found. Does this correlation imply that cell wall changes are causal to morphological changes or vice versa? Many workers favor the first explanation for the following reasons: (1) Mutants known to be blocked in particular steps in cell wall synthesis have drastically altered morphologies; (2) all the morphological mutants studied to date, with only one exception, have a gross cell wall composition different from the wild type (DeTerra and Tatum, 1963; Mahadevan and Tatum, 1965); and (3) fungal protoplasts usually have a spherical configuration, suggesting that the shape of the cell is primarily determined by the cell wall, and that intracellular structures play a minor role. Therefore, a reasonable working hypothesis is that the cell walls of fungi are the key structures that determine cellular shape and colony pattern formation.

The biosynthesis of fungal cell walls have not been studied in as much detail as the biosynthesis of the bacterial cell walls. In many cases, only the gross chemical composition or the types and amounts of different polymeric constituents are known. In a few cases, the precursors to the homopolymers are known to be nucleotide sugars (Glaser and Brown, 1957). The polymerization reactions are catalyzed by particulate enzymes which appear to be located inside the cell membrane. However, there are reports of this type of enzyme being associated with the cell wall (Wang and Bartnicki-Garcia, 1966; McMorrough et al., 1971). Whether these associations reflect localization in the cell wall or a high affinity for some cell wall component after disruption of the cell is still an open question. In the case of chitin synthetase of yeast the enzyme is clearly not outside the cell, since regenerating protoplasts are insensitive to the specific inhibitor (polyoxin) of this enzyme (Cabib and Keller, 1971).

Little definitive information is available about the mechanics and chemistry of the secretion and assembly of polymers into the cell wall. Reports of specialized secretory vesicles (lomasomes) and other bodies (Spitzenkörper) near the apex of growing hyphae of certain fungi (Grove and Bracker, 1970; McClure et al., 1968; Girbardt, 1969) are suggestive but not conclusive. No clear chemical demonstration as to which *specific* precursor materials are present in these vesicles has been reported so far. The assembly of fibrils into the cell wall has not been investigated in any detail either. The cell wall degrading enzymes associated with cell walls (Mahadevan and Mahadkar, 1970; Mahadevan and Menon, 1968) may play a role in intercalating new fibrils into the existing matrix or a role in cell wall turnover. One phenomenon which may be related to the assembly of the β-$(1 \rightarrow 3)$-glucan into the *Neurospora* cell wall is the effect of sorbose on colony morphology (DeTerra and Tatum, 1961; Tatum et al., 1949). When a high concentration of sorbose is added to the medium, the wild-type strain grows with a tight colonial morphology. It is known that under these conditions very little sorbose is found inside the cells (Crocken and Tatum, 1967) and that the level of β-$(1 \rightarrow 3)$-glucan in the cell wall is greatly reduced (Mahadevan and Tatum, 1965). It has also been reported that sorbose affects an extracellular polysaccharide degrading enzyme in *Schizophyllum* (Wilson and Niederprem, 1967b) and that the addition of a mixture of hydrolytic enzymes to the medium reverses the sorbose effect in *Neurospora* (Rizvi and Robertson, 1965). Based on these four observations, one can postulate that sorbose inhibits an enzymatic process involved in the extracellular assembly of glucan fibrils into the cell wall and that this inhibition produces an altered phenotype.

The role of each component in determining the rigidity and plasticity of a given cell wall is also not well documented and may vary from organism to organism. Cell shape changes in *Mucor rouxii* are correlated with changes in amount of one of the cell wall polysaccharides, a mannan (Bartnicki-Garcia, 1968; Bartnicki-Garcia and Nickerson, 1962). Unfortunately, it is not clear whether these mannan changes are the cause of the altered morphology or whether they reflect some other changes in cell wall composition. Studies on the chitin content of temperature-sensitive *Aspergillus* mutants (Katz and Rosenberger, 1970) indicate the relative unimportance of a severe deficiency in this component in overall shape determination, but the importance of this component in cell wall rigidity. Similarly, the addition to the media of a specific *in vitro* inhibitor of chitin synthetase stopped the growth of *Neurospora*, but did not produce morphological variations prior to that point (Endo *et al.*, 1970). On the other hand, the β-$(1 \to 3)$-glucan found in *Neurospora* cell wall appears to be important for morphogenesis, since mutants partially blocked in the synthesis of this polymer have grossly altered phenotypes (Brody and Tatum, 1967b). At the present time, it appears that the level of almost every component of the *Neurospora* cell wall can be individually altered by mutation with various consequences to morphogenesis. The full extent of this variation remains to be explored, since some cell wall mutants may require special selective conditions, such as high osmotic strength, for isolation.

The localization of any individual component within a fungal cell wall is a somewhat controversial subject and may differ from one organism to another (Hunsley and Burnett, 1970; Troy and Koffler, 1969). In general, all the different components appear to be intermeshed with each other and one cannot visually observe, in a cross-sectional view, any clear-cut spatial distinction between layers (Manocha and Colvin, 1967). Therefore, models of the cell wall architecture primarily rely on the results obtained from sequential digestion of the components, as well as some microscopic observations. This type of analysis has led some to propose that certain fungi have a chitinous inner layer covered by a glucan layer (Livingston, 1969: Mahadevan and Tatum, 1967), while others favor a covering layer of protein fibrils (Hunsley and Burnett, 1970). Since these models are based on either chemical or enzymatic techniques, perhaps the use of a different methodology, such as a mutational approach, might settle the controversy.

D. Relation between Cell Wall Changes and
   Morphological Changes in Fungi

The chemical description of cell wall changes is a very different type of description than that used to analyze the growth pattern of mycelia.

The pattern of a growing fungal colony is usually described macroscopically and physically by the folowing characteristics:

(1) The apical growth rate of the main hyphae and secondary hyphae

(2) The frequency of branching per unit length and per unit time

(3) The angle of branching from the main hyphae

(4) The diameter of the main and secondary hyphae

(5) The size and shape of individual cell compartments, i.e., bulbous vs. cylindrical

(6) The compactness of growth, expressed as mycelial mass per unit area.

These six measurements are esentially performed on a two-dimensional mycelial network (Butler, 1961). The three-dimensional aspect of colony morphology is far more difficult to analyze and cannot always be correlated with the two-dimensional analysis (Trinci, 1969). Another phenomenon which affects the shape of a colony, but cannot be quantitatively measured, is hyphal tip–tip interactions. For example, the growth of one hyphal tip in front of another hyphal tip in the wild-type *Neurospora* is associated with the cessation of growth of the latter hyphae. This phenomenon is not found in some of the morphological mutants of *Neurospora* or in the wild type of *Aspergillus* (Weston, 1965) in which growing tips continue to grow and branch in the interior of the colony, thus leading to a very compact and often hemispheric growth. But even if this interaction could be quantitated along with the six measurements above, there still exists a basic dichotomy between the description of an event in terms of its visible parameters and in terms of the associated chemical changes.

These two modes of analysis can be brought to bear upon the principal and critical question in vegetative morphogenesis: how do branches form? One would like to know: How and where branches are formed; if there are special sets of enzymes needed for branch formation; if there are many weak parts or special parts of the cell wall where branches can be formed (*i.e.*, apical vs. subapical branches); why the distance between branches is fairly constant; and what the mechanism is of orientation of branching (right vs. left side). Questions may also be asked as to the inverse relation found between apical growth and branching. That is, does a lower apical growth rate produced by a mutation lead to an increased branch frequency or do these mutations increase the branch frequency and therefore decrease the apical growth rate? In addition, one would like to know about the plasticity of the cell wall and the turgor pressure (Park and Robinson, 1966; Robertson, 1968) involved in advancing the mycelial front. Our understanding of all these questions is very meager at this time, but a start has been made. For example, the bud–mycelial form transition of *Mucor rouxii* is associated with a diffuse (bud) vs. localized (mycelial) pattern of incorporation of cell wall precursors into cell wall (Bartnicki-Garcia and Lippmann, 1969). In *Schizophyllum*, induced morphological variants were found to have varied ratios of a resistant (R) glucan to a soluble (S) glucan, and there is a consistent correlation between low levels of a particular glucan and increased levels of a specific R-glucanase (Wilson and Niederpruem, 1967a). At the present time, it is not clear whether the change in glucanase level is the cause of the altered morphology or the result of an altered metabolism. Another interesting report shows the localization of chitin almost exclusively to the bud scar region of yeast (Bacon *et al.*, 1969), and suggests a role for this component in providing a rigid wall during the budding process (Cabib and Bowers, 1971). In summary, a reasonable amount of information has been gathered about cell wall composition and some information about the kinetics of fungal growth, but little is known relating the two. Section IV below reports on the rudimentary knowledge of these problems in *Neurospora*, as well as a brief description of other areas of *Neurospora* morphogenesis.

E. Other Aspects of Fungal Morphogenesis

The preceding paragraphs were focused on aspects of vegetative morphology and their relation to the cell wall. In fungi, this process requires growth and increase in mass. However, many other fundamental, developmental phenomena in fungi, such as conidiation and sporulation, do not require active growth, but occur under conditions of general starvation. This observation may be extended to other microbial processes, such as the sporulation of certain bacteria (see Section III,B), the morphogenesis of cellular slime molds (Sussman, 1967), and the formation of flagella in *Naeglaria* (Dingle and Fulton, 1966). This type of catabolic or "last gasp" differentiation may emphasize different metabolic pathways than those pathways required for active proliferation. In fact, interference with the functioning of the Krebs cycle appears to affect the catabolic differentiation of *Neurospora* far more severely than vegetative morphogenesis (Ojha *et al.*, 1969). These differential inhibitions have not yet been firmly traced to different sets of isoenzymes, but there are some interesting preliminary reports (Peduzzi and Turian, 1969).

There are other structures formed by fungi which require some growth and increase in mass, but are probably formed in response to some type of specific starvation conditions, such as nitrogen limitation. These include the female receptor bodies (protoperithecia) which are formed during the sexual cycle. The physiology and development of these structures have been investigated (Barbesgaard and Wagner, 1959; McNelly-Ingle and Frost, 1965; Rothschild and Suskind, 1966), but many preliminary descriptive details about the chemical composition of these structures are still lacking. In fact, the physiology of the entire meiotic process in fungi has been analyzed far less than the vegetative part of the life cycle, probably owing to difficulties in obtaining sufficient quantities of material. However, many of the processes in the sexual cycle appear to utilize the same enzymes as the vegetative cycle, as exemplified by the effects of certain point mutations on both phases of the *Neurospora* life cycle. These observations and others are pointed out in the following pages (Section IV).

For a general overview of fungal morphogenesis, one can find ample material in the review by Morton (1967), a monograph by Turian (1969), the unusually well-documented text by Esser and Kuenen (1967), or in the multi-volume treatise, "The Fungi" (Ainsworth and Sussman, 1966). Selected aspects of morphogenesis of *Neurospora* have been covered by various workers, such as Turian (in Ainsworth and Sussman, 1966), Robertson (1968), Zalokar (1959), and Barber *et al.* (1969). Articles on other fungi can be found for *Phycomyces* by Bergman *et al.* (1969), for *Podospora* by Lysek and Esser (1971), for *Paracoccidioides* by Kanetsuna and Carbonell (1966), for *Blastocladiella* by Cantino and Lovett (1964), and for *Schizophyllum* by Niederpruem and Wessels (1969).

IV. Specific Experimental Information about *Neurospora* Morphogenesis

A. Genetic Knowledge

The genetics of *Neurospora crassa* have been well studied, both intensively and extensively. The total number of known loci is approximately 400, and these are distributed fairly evenly over the seven linkage groups. Approximately 30% of these loci are associated with morphologically distinct phenotypes. This may be a minimum estimate for a variety of reasons. Many mutations do not lead to changes dramatically obvious to the naked eye, and most investigators do not microscopically examine the survivors of a mutant hunt. In addition, morphological mutants have rarely been the object of specific selection experiments, but have usually been isolated as interesting oddities that appeared during auxotroph hunts. Notable exceptions are the isolation of conidiation mutants by R. W. Siegel (personal communication), ascus mutants by Srb and Basl (1969), and general morphological mutants by Perkins *et al.* (1969) and Dicker *et al.* (1970). Third, many morphological mutants have not yet been mapped, so it is not clear if they represent new loci. Finally, almost all the existing morphological mutants are prototrophic, which is not surprising since they were isolated on minimal media. Whether a class of undetected morphological mutants exists that has additional growth requirements is not clear at this time. Perhaps the percent and numbers of morphological mutants may rise somewhat when the proper techniques of selection, isolation, and cultivation of these mutants are performed.

The morphological mutants already described show little clustering on the genetic map. The closest linkage relationships described to date for a group of morphological mutants of the same phenotype are the three *crisp* loci; approximately 10–15% recombination units separate these loci (Garnjobst and Tatum, 1970). In general, mutants with the same phenotype are unlinked to each other, such as the *ropy* mutants, which are located at seven different sites in five different linkage groups (Garnjobst and Tatum, 1967). Occasionally, morphological mutants of very different phenotypes map near each other, but it is unclear if they are biochemically related to each other.

Most of the morphological mutations described so far appear to be point mutations in that they can be easily mapped, do not change nearby linkage relationships, and revert readily to the wild type. Most of them are recessive in heterokaryon tests, while the exceptions are not strictly dominant either. Interestingly, certain morphological mutations appear to act as recessive in heterokaryon tests but act as dominants for ascus morphology (Novak and Srb, 1971b). This is a clear illustration of the differences in the genetic determination of metabolism in the asexual and sexual phases of the *Neurospora* life cycle. Two additional facets of the genetics of morphological mutants are that double mutants generally have more extreme phenotypes than either single mutant strains (Garnjobst and Tatum, 1967; Mahadevan and Tatum, 1967) and that genetic modifiers are known for many of the mutant alleles. These modifiers may either increase or decrease the growth rate of a given morphological mutant.

The existing morphological mutants can be divided into at least five classes with respect to their effects on different phases of the life cycle. The first class consists of those mutants in which the branching pattern of the vegetative mycelia is affected. This class includes the slow growing, very compact, and highly branched mutants called colonial mutants, as well as other vegetative morphological variants (Garnjobst and

Tatum, 1967). The circadian rhythm mutants should also be included in this group, since their phenotype is caused by periodic increases and decreases in branching frequency (Sargent *et al.*, 1966; Sussman *et al.*, 1964). So far, most of the known morphological mutants of *Neurospora* fall into this first group with no particular distribution of types. There are 60–80 mutant loci in this group, and one-third of these were detected in the past few years. Although it is not surprising that repeat mutations at certain loci have been found, it is surprising that these repeat mutations usually have the same phenotype as the original isolate. This is a puzzling observation, since one would expect slightly different phenotypes to be caused by different allelic forms of the same locus. Hopefully, this puzzle will be solved when more mutants have been isolated and individual loci become saturated with mutations.

The second class of mutants are those that are unable to form distinct macroconidia but are otherwise normal in the vegetative branching pattern. Some of these strains have been known for some time (Sheng and Ryan, 1948), and recently some ingenious techniques have been developed for the specific selection of aconidial (*acon*) mutants (Siegel *et al.*, 1968). There are at least five separate loci involved in the formation of macroconidia (R. W. Siegel personal communication), and mutants at each of these loci are blocked at a different stage in conidia formation. Mutations affecting other aspects of conidia formation have been detected previously (Grigg, 1960a,b). It should be pointed out that many of the extreme colonials do not form conidia either, but this is probably a secondary effect due to the inability of the abnormal, bulbous-shaped cells to pinch off properly into conidia. It is clear that there is a class of genes which are involved solely with the formation of these vegetative spores (conidia).

Another class of mutants, the germination mutants, can be further subdivided into conidial germination mutants (Inoue and Ishikawa, 1970), ascospore nongerminators and ascospore spontaneous germinators (Emerson, 1963). Some of the mutations which lead to these phenotypes may have additional phenotypic manifestations in other phases of the *Neurospora* life cycle. A fourth class of mutants are the sterility mutants, those that either cannot form visible protoperithecia (Tee and Choke, 1970) or form some type of altered but sterile female receptor bodies. Some of these have no obvious alterations in their vegetative state. A fifth type of mutant is altered in the postfertilization events, such as the formation of the proper number of spores in the ascus (Novak and Srb, 1971a), their linear arrangement (Pincheira and Srb, 1969a,b), or the morphology of the individual ascospores (Leary and Srb, 1969). Here again, many mutant strains are known which are affected in both vegetative morphology and ascospore formation, indicating common biochemical processes. This is to be expected since these mutants were originally screened and isolated on the basis of their vegetative morphology (Srb and Basl, 1969). These several classes of mutants represent a wealth of biological material, and it is expected that the combination of novel selective techniques with the use of conditional (temperature-sensitive) mutants will lead to the isolation of phase-specific mutants, as well as some more general conclusions about the percent of morphological mutant loci, their clustering, and the relative size of these five classes.

B. PRIMARY ENZYME DEFICIENCIES

The primary enzyme deficiencies have been elucidated for eleven of the 100–120 morphological mutants described to date. These eleven mutations affect carbohydrate or phospholipid metabolism. They are listed below along with their locus designations (italicized):

1.	glucose-6-phosphate dehydrogenase	*col-2*
2.	glucose-6-phosphate dehydrogenase	*bal*
3.	glucose-6-phosphate dehydrogenase	*frost*
4.	6-phosphogluconic acid dehydrogenase	*col-3*
5.	6-phosphogluconic acid dehydrogenase	*col-10*
6.	phosphoglucomutase	*rg-1*
7.	phosphoglucomutase	*rg-2*
8.	phosphohexoseisomerase	?
9.	inositol phosphate synthetase	*inos*
10.	phosphatidylmonomethylethanolamine *N*-methyltransferase	*chol-2*
11.	phosphatidylethanolamine *N*-methyltransferase	*chol-1*

The last three strains (9–11) on this list are morphologically aberrant only when grown on suboptimal levels of inositol (Pina and Tatum, 1967) or choline (Crocken and Nyc, 1964; Scarborough and Nyc, 1967), respectively. When grown on high levels of the supplements, they have normal morphology. These three mutants are the only well-documented examples in which auxotrophic mutations have morphological effects under certain conditions. All the other types of auxotrophs, such as those that require amino acids, purines, or vitamins will only grow slowly and sparsely under conditions of suboptimal supplementation and do not grow as compact, highly branched colonies. The chain of events produced by the low supplementation of inositol or choline leading to the abnormal morphology is not known, but it has been reported that under these conditions that the membranes of the *inos* mutant have approximately one-fifth the wild-type level of phosphatidylinositol (Fuller and Tatum, 1956; Germanier, 1959). When *inos⁻* is allowed to grow further and use up all of the inositol in the medium, death occurs (Lester and Gross, 1959; Matile, 1966). But prior to that point, the strain grows as a compact colonial, and preliminary data (S. Brody, unpublished) indicates that there are differences from the wild type in the composition of the cell wall. In these respects the *inos⁻* of *Neurospora* is similar to the inositol-requiring yeast (Ghosh *et al.*, 1960; Power and Challinor, 1969). Precisely what step or steps in the biosynthesis of the cell wall has been disrupted by inositol deprivation is not known. Possibly the transport and assembly of cell wall components are defective. The two choline-requiring mutants show a similar response to low choline in terms of colonial morphology, abnormal phospholipid composition, and eventual death (Crocken and Nyc, 1964). In addition, the uptake of niacin has been impaired (Lie and Nyc, 1962). Preliminary evidence also indicates that the cell wall changes when the strain is grown in the colonial form (S. Brody, unpublished). Both of these types of mutants are potentially very useful in anayzing the structure–function relationships of individual membrane components.

The elucidation of the primary enzyme defect is relatively straightforward for any auxotroph that has only a single requirement, such as the inositol- or choline-requiring strains. However, the same type of analysis for mutants isolated on the basis of aberrant morphology is not at all straightforward, and usually cannot be corroborated by a phenotypic reversal due to the addition of a single compound. Therefore, a combination of screening procedures, stringent analyses of purified enzymes, and some *in vivo* measurements must be employed. Using these procedures, the primary enzyme defect has been pinpointed in eight out of the 40–50 mutants that have been screened in some detail.

The first to be elucidated was *col-2*, originally isolated in 1949 (Barratt and Garnjobst, 1949) and described as a glucose-6-phosphate dehydrogenase (G-6-PD) mutant in 1966 (Brody and Tatum, 1966). This strain accumulates G-6-P and possesses normal levels of a G-6-PD of altered thermal stability and poorer substrate affinity (Brody and Tatum, 1966). Further analysis indicated striking differences between the patterns obtained from isoelectric focusing techniques of homogenous G-6-PD preparations of *col-2* vs. wild type (Scott and Tatum, 1970). Experiments with revertant strains indicated that mutations at the *col-2* locus which restored the wild-type morphology simultaneously restored the G-6-PD properties of the wild type. Temperature-sensitive partial revertant strains of intermediate morphology possess a G-6-PD which is temperature-sensitive and has properties intermediate between those of the wild-type and *col-2* enzymes (Brody and Tatum, 1966). Recently, an unlinked specific suppressor gene of *col-2* was detected which changes the morphological phenotype of *col-2* (S. Brody, unpublished) as well as the properties of the G-6-PD (W. A. Scott, personal communication). All these observations implicate the *col-2* locus as a structural gene for G-6-PD. A discussion of the pleiotropic effects produced by alterations in G-6-PD is given in Section IV,C below.

Other morphological mutations at the *bal* and *frost* loci also lead to alterations in G-6-PD (Scott and Tatum, 1970). The mutations are unlinked to each other or to *col-2*. No partial revertant strains or specific suppressor strains of these mutants have been described so far, but a double mutant (*bal, col-2*) has a very slow growth rate and a very labile G-6-PD activity. In addition, heterokaryons formed between *bal* and *col-2* strains show an unusual isozyme pattern (Scott and Tatum, 1970). These facts would appear to indicate that *bal* is also a structural locus for G-6-PD and that all three mutants have a similar primary enzyme defect in the first step in the pentose phosphate shunt.

Mutational alterations in 6-phosphogluconic acid dehydrogenase (6-PGAD), the next enzyme in the pentose phosphate shunt, also lead to morphological aberrations. These mutations occur at two unlinked loci, *col-3* and *col-10*. The enzymatic evidence is of a similar type to that reported above and consists of altered affinities and changes in isoenzyme patterns (Lechner and Fuscaldo, 1969; Abramsky *et al.*, 1971). As in the examples of *col-2* and *bal*, these two 6-PGAD mutants are slow growing, compact colonials and contain the same specific activity of the altered enzyme in crude extracts as the wild type. However, no substrate accumulation (6-PGA) has been reported and no measurements of the *in vivo* rate of the shunt have been made on these two mutants. It is interesting that in all five cases, the mutational alteration does not appear to affect the level of the altered enzyme when assayed with substrate excess, but does have dramatic effects on the cellular metabolism and morphology. This indicates both the critical role of an enzyme's substrate affinity under *in·vivo* conditions (substrate limiting) as well as the importance of the products of the pentose phosphate shunt. Whether the affinities of these enzymes are rate limiting in wild type and therefore play a significant role in the control of normal morphogenesis is an open question at this time.

In contrast to the five separate mutations listed above, the morphological mutations at the two *ragged* (*rg*) loci which affect the structure of phosphoglucomutase (PGM) lower the levels of the enzyme activity in crude extracts (Brody and Tatum, 1967b; Mishra and Tatum, 1970)

to only 10–15% of the wild-type activity. These may be minimum esti-
mates since no serious efforts have been made toward changing the assay
conditions to increase these basal levels. The assayable enzyme activity
in these mutants is quite unstable and when partially purified shows
pronounced electrophoretic differences from the wild type (Mishra and
Tatum, 1970). No partial or complete revertant strains have been isolated from these mutants, although a suppressor mutation has been described (Mishra and Threlkeld, 1967). Some puzzling features of the
rg mutants are described in Section IV,C.

The most recent report of an enzyme defect which leads to morphological changes is that of an altered phosphohexoseisomerase (Murayama,
1969). Very few details are available, but a translation of that report
indicates that growth on glucose leads to an altered morphology and
slow growth, and that growth on glucose and fructose reverses the morphological effects. Cell wall differences were reported for this mutant
when grown on glucose, but it was not indicated which cell wall fractions
are altered. One might suspect a deficiency in chitin formation, since
the first reaction in that pathway utilizes fructose-6-phosphate.

It is curious that so far all eight of these mutants have alterations
in the central area of metabolism involving glucose-6-phosphate. It is
to be expected that the primary deficiency in the next eight mutants
to be reported will probably be more specifically involved with cell wall
formation, since the screening of mutants is now focused on cell wall
precursor synthesis, polymerization, and assembly reactions. However,
there is something unusual about the altered central metabolism of these
first eight mutants which leads to shape and pattern changes. Other
mutations which affect central areas of metabolism, such as the Krebs
cycle or the gloxylate shunt, do not affect vegetative morphology. For
example, glutamic dehydrogenase, malic dehydrogenase, succinic dehydrogenase, pyruvic carboxykinase, isocitrate lyase, malate synthetase,
and fatty acid synthetase mutants are all without morphological consequences. The differences in pleiotropic effects produced by these two
groups of mutants of central metabolism are not clear, and, therefore,
only the first group is discussed in the following paragraphs.

C. PLEIOTROPIC EFFECTS

The three G-6-PD mutants, *col-2*, *bal*, and *frost*, all have certain
basic features in common when compared to the wild-type strain. They
all accumulate the substrate G-6-P and show lower levels of reduced
nicotinamide adenine nucleotide phosphate (NADPH), a product of
G-6-PD (Brody, 1970b). They all grow as tight, compact colonials in
liquid shake cultures as compared with the spreading filamentous mat

of the wild type. They all show lower levels of linolenic acid in their membrane lipids (Brody and Nyc, 1970). They all show lower levels of all four pyridine nucleotides, not just NADPH.

The differences among these three mutants are probably more interesting, particularly the readily observable differences in morphology (see Fig. 1). The *col-2* and *balloon* mutants appear very similar when grown on solid media, although the *col-2* strain has a bumpy surface and the *balloon* strain a very smooth hemispheric shape. Time-lapse photography of germinating ascospores of *bal* or *col-2* from selfed crosses indicate a significant difference in the growth pattern of these two mutants (S. Martins, personal communication). The mycelia of the *col-2* strain start to branch immediately upon germination and continue growing that way, whereas the *bal* strain has a short period of relatively linear apical growth with no branching followed by periods of intensive branching. The third mutant, *frost*, grows considerably faster on agar than either *bal* or *col-2* and, in addition, shows a unique growth pattern. Hyphae on the surface of a solid medium branch extensively and therefore have a slow apical growth rate, whereas aerial hyphae protrude at an angle to the surface and grow quickly with little branching for a considerable length. Eventually, these aerials branch somewhat and fall to the surface of the agar where they continue to branch profusely. This pattern of growth accounts for the frosty weblike appearance of the colony on a plate and for its relatively rapid increase in colony diameter, which is approximately two-thirds the wild-type rate and twentyfold greater than *bal* or *col-2*. This bizarre pattern is changed when the *frost* mutant is incubated upside down, where it has a much slower growth rate and more compact colonial morphology (Weston, 1965), although still significantly different from *bal* or *col-2*. When grown as a liquid shake culture, it has a semicolonial morphology and a growth rate intermediate between wild type and *col-2*. In summary, these three mutants display morphological differences under all growth conditions.

These three mutants also show distinct biochemical differences, as discussed below, although the most detailed studies have been done on *col-2*. Radiorespirometry measurements performed on *col-2* indicate an approximate 50% reduction in the *in vivo* activity of the pentose phosphate shunt (Brody and Tatum, 1967a). Similar analyses have not been performed for *bal, frost,* or the revertants and modifiers of *col-2*, but it is expected that these strains will also show lower levels of this pathway. In each case, the question can be posed as to whether a postulated decreased *in vivo* activity leads to lower levels of NADPH and/or pentose phosphates. Growth of all three mutants on xylose or ribose as sole carbon sources or as auxiliary carbon sources does not alter the morphology or growth rate appreciably, thereby suggesting that their

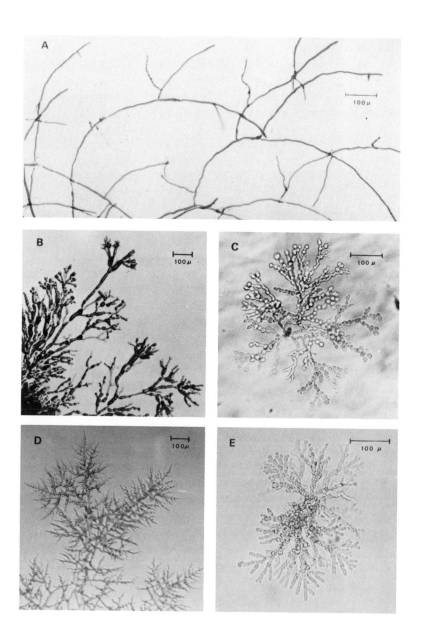

FIG. 1. Photomicrographs of *Neurospora crassa* grown on agar media at 24°C.
A. Edge of a wild-colony. B. *Frost* mutant (G-6-PD). C. *Col-2* mutant (G-6-PD).
D. *Bal* mutant (G-6-PD). E. *Rg* mutant (PGM).

altered metabolism is not principally due to a defect in the production of pentose phosphates (Brody and Tatum, 1967a). Therefore, attention has been focused on the decreased NADPH levels in all these strains as the more critical deficiency caused by these G-6-PD mutations.

There is a general correlation between the growth rate and the total NADPH content for *col-2; bal; frost; col-2, su;* and the *col-2* revertant strains. There is also an excellent correlation between the total NADPH content and the percent of linolenic acid in the lipids of *Neurospora* (Brody and Nyc, 1970). As one might expect, there is also a fine correlation between the percent of linolenic acid and the growth rate of these strains. The finding that all three of these parameters vary simultaneously and linearly indicates that all three are in some way related to a single cause (*i.e.*, a change in G-6-PD). However, a clear understanding stops at that point since mutants, such as *bal* and *col-2*, with the same low NADPH content have distinctly different morphologies. One is forced then to consider the role of each of the isoenzymes of G-6-PD in the cell's metabolism and/or the possible compartmentalization of the NADPH produced by each of these isoenzymes.

One possible explanation would be that the individual polypeptides that are thought to make up the G-6-PD in *Neurospora* are also subunits for some other enzyme or enzymes. A related possiblity would be that an entire G-6-PD isoenzyme, not just one of its subunits, is part of some other enzyme aggregate and that each of the G-6-PD isoenzymes is necessary for the activity of some different aggregates. In either case, one mutation would produce more than one primary effect.

A third possibility involves the localization of these isoenzymes and/or the compartmentalization of one of their products (NADPH). It can be proposed that the NADPH produced by one G-6-PD isoenzyme is more directly utilized for fatty acid biosynthesis, while the activity of another G-6-PD isoenzyme is coupled with ergosterol biosynthesis. If each isozyme produces a separate pool of NADPH, then the mutation at the *bal* locus could severely deplete one NADPH pool, and the *col-2* mutation another. If one accepts these assumptions of coupled reactions and loss of specific NADPH pools, then one could get an unequal effect on certain cellular reactions and ultimately different morphologies. It should always be kept in mind that the same low level of NADPH in both *bal* and *col-2* reflects a decrease in *total* content and does not indicate the severity of the deficiency for any given enzymatic reaction or for any given location in the cell. One has the additional problem in *Neurospora* of mycelial localization: Is one of the isoenzymes present only at or near the tip of the growing mycelia?

In order to understand how these presumed single polypeptide changes lead to the different phenotypes, an investigation of the pleiotropic effects of these mutations was undertaken. As pointed out above, the first principal effect of the G-6-PD deficiency is a decreased NADPH content. But it was also found that there are decreased levels of other pyridine nucleotides as well (Brody, 1970b). Our explanation is that the other coenzymes appear to be decreased due to the presence of a positive feedback loop in pyridine nucleotide synthesis in *Neurospora*. It is known that in *Neurospora*, NADPH is required for the activity of kynurenine hydroxylase, a step in the synthesis of pyridine nucleotides (Cassady and Wagner, 1971). Therefore, it is possible that a decrease in NADPH could lead to a lower rate of pyridine nucleotide synthesis, which would lead to even lower NADPH levels. This may very well be the case since the addition of nicotinic acid to the medium, which enters the pathway past kynurenine hydroxylase, raises somewhat the NAD and NADH levels in *col-2*, but not the NADP or NADPH levels (Brody, 1972). This partial reversal of one of the pleiotropic effects does not change the morphology of *col-2*, indicating that it is the NADPH level which is the cause of the aberrant phenotype.

Another pleiotropic effect produced by the lowered NADPH content is the low level of linolenic acid in the neutral and phospholipids (Brody and Nyc, 1970) in all three G-6-PD mutants. This is interpreted as a specific *in vivo* NADPH requirement for the final desaturation step in linolenic acid biosynthesis. Another effect, assayed only in *col-2* so far, is a lower level of reduced glutathione which would implicate NADPH in the enzymatic reduction of glutathione. A third pleiotropic effect is a large accumulation of neutral lipids in *col-2*. This property appears to be unique to *col-2* since it is not found in *bal*, *frost*, or *col-3*, all mutants with low NADPH levels (Nyc and Brody, 1971). This finding suggests unique effects due to the mutations in the different isoenzymes of G-6-PD. Another finding consistent with this idea is that the substitution of glutamic acid for glucose as the sole carbon source partially reverses the *col-2* morphology, probably by generating more NADPH via the NADP-specific glutamic dehydrogenase. However, the *bal*, *col-3*, or *frost* strains were not remedied by growth on glutamate. These differences between the three G-6-PD mutants might provide some clues as to the metabolic significance of the individual G-6-PD isoenzymes or their subunits. Perhaps the different morphological phenotypes are due to a combination of a NADPH deficiency with these other secondary effects.

No discussion of coenzyme levels and isoenzyme localization is complete without some reference to the compartmentalization of metabolites. Some of the best evidence about pools of intermediates and their relation-

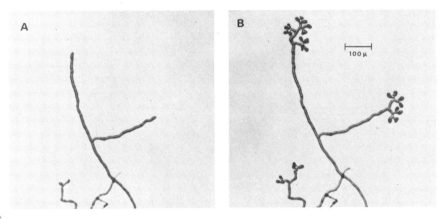

FIG. 2. The effects of a temperature shift on a strain of *Neurospora* with a temperature-sensitive G-6-PD. A. Growth on agar media at 24°C. B. The initiation of multiple branches 1 hr after temperature shift to 34°C.

ship to each other has been obtained with *Neurospora* (Davis, 1967). Although the physical basis for the separation of pools in *Neurospora* is not clear, some new evidence has been obtained regarding NADPH levels. Employing temperature-sensitive partial revertants of *col-2*, we have noticed dramatic changes (see Fig. 2) in the morphology of the growing tips shortly after a 10°C-increase in the temperature. This startling increase in apical branching is followed by continuous branching and compact colonial growth as long as the temperature remains at 34°C (Brody, 1970a). Measurements of the levels of pyridine nucleotides indicate that their 34°C steady state level is not reached until 20–24 hours after the temperature shift; the levels do not decrease suddenly following the increase in temperature. In fact, after one hour at 34°C, only an 8–10% decrease in NADPH content is found, yet the morphology at the tips is strikingly different! At least two interpretations of this data are possible: (1) The NADPH content everywhere in the mycelia is reduced slightly and the growth of a hyphal tip is extremely sensitive to this small reduction and branches profusely and (2) the NADPH pool of the tip regions is quickly and severely depleted, causing branching to occur. If the NADPH pool at the tip is only a small fraction of the total NADPH content, then one would expect only an 8–10% initial decrease in the total content (Brody, 1970a). The second explanation is favored at the moment and definitive experiments are underway to distinguish between these two alternatives.

The pleiotropic effects produced by a G-6-PD deficiency and a NADPH deficiency can be numerous and far reaching (see Fig. 3).

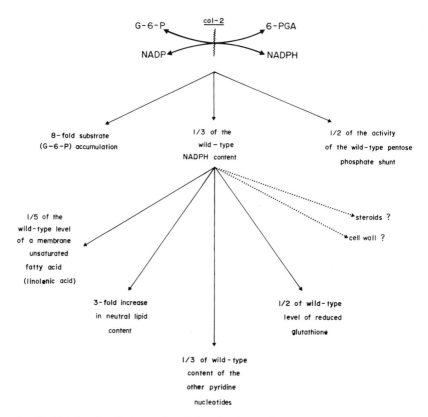

FIG. 3. Pleiotropic effects produced *in vivo* by a G-6-PD mutation. For details, discussion, and citations, see text.

In one sense, this type of mutation may not be the best type to use as a tool for dissecting the relation between enzymes, cell walls, and morphology since our studies of G-6-PD mutants can be interpreted to indicate a general debilitating effect of a mutation in central metabolism. Clearly, those mutations which affect an enzyme directly involved in cell wall synthesis and which are unique to cell wall synthesis offer more opportunities for a straightforward and uncomplicated analysis. In this regard, it is curious that the *col-2* strain is the only one of the many morphological mutants analyzed that does not show striking changes in the gross composition of the cell wall. One interpretation of this observation is that a minor component of the cell wall has been altered in *col-2*, but has not been detected so far. Another possibility is that the rate of secretion of many cell wall components has been simultaneously reduced in this strain and that this limitation in general cell wall synthesis may lead to increased branching and bulbous cells. At the moment, there are no observations to support either of these two postulates.

The pleiotropic effects produced by the *ragged* mutations are, in part, somewhat easier to interpret than those of *col-2*. The severe block in phosphoglucomutase (PGM) activity (G-6-P ↔ G-1-P) in this strain leads to a 50–60% reduction in the levels of a β-(1 → 3)-glucan in the cell wall, and an even greater decrease in the glycogen content of the mycelia. This is not unexpected, since it appears that uridine diphosphoglucose (UDPG) serves as a precursor for both of these polymers and that UDPG is formed in *Neurospora* by the reaction G-1-P + UTP ↔ UDPG. However, what is still puzzling about these *rg* mutants is the fact that they have normal levels of UDPG and that they accumulate G-1-P. One would expect lower levels of all intermediates in a pathway which was blocked at the first step, not normal and higher levels. In addition, the G-1-P level of *rg* does not appear to be in complete equilibrium with the G-6-P pool, based on known equilibrium ratios of the enzyme and on the G-1-P:G-6-P ratio in the wild-type strain. A fuller discussion of these points is given in the original paper (Brody and Tatum, 1967b), but it should be pointed out that one possible explanation would be that the subunits of PGM may be involved with the functioning of some other enzyme as well. Some additional data on the G-6-P pool size were obtained when it was possible to isolate the very slow growing double mutant (*rg, col-2*) from ordered asci. This double mutant strain has the properties of both of its parents, *i.e.*, low NADPH, low β-(1 → 3)-glucan, and it accumulates as much G-6-P as the *col-2* strain (Brody, 1969). Therefore, it would appear that increased levels of G-6-P in a strain partially deficient in PGM does not divert more G-6-P into the cell wall, as measured by the level of β-(1 → 3)-glucan.

The tracing of the pleiotropic effects produced by a mutation is a difficult and convoluted task, even when the primary enzyme deficiency is known. However, there are two approaches which show some promise in helping with this undertaking. The first approach, as indicated in Section II above, is the reversal of the mutant phenotype by the addition of a permeable intermediate. This can give very definite metabolic clues, even though they may sometimes be indirect. The second approach is the isolation of revertants with more normal phenotypes, particularly those due to unlinked suppressor mutations. Disregarding the type of suppressor gene which repairs the altered protein via mistakes in protein synthesis (Brody and Yanofsky, 1963), one should concentrate on those unlinked mutational events that change the metabolism in some compensatory manner. For example, if a pathway was thought to be limiting due to the pleiotropic effects of a lowered steady state level of coenzyme or substrate, then one might try to isolate strains containing derepressed levels of this pathway. In this way, the organism might provide the identification as to which particular pleiotropic effect was causal to the altered morphogenesis and which effects were peripheral.

V. Outlook

Our present knowledge of cell walls and morphogenesis in fungi can be roughly classified into four categories. The first category, the enzymes and substrates involved in the biosynthesis of cell wall components, has been partially described by many workers (see Section III,C) in a relatively straightforward way. The second category, the mechanics and chemistry of assembly of individual cell wall components into the cell wall, has not been well described. The third category, the relationship between observed cell wall changes and the observed visible changes in branching and cell shape, consists of a few scattered observations. The fourth category, the relation between changes in branching frequency or cell shape and the gross morphology of an entire colony, has been described in some detail (Section III,D). So, it is clear that the missing links are the details about the coordination of synthesis and assembly of cell walls, and the effect of regulatory perturbations on the kinetics of apical growth and branching. A great deal of information about these missing links could be gathered by studying how the synthesis of an enzymatically inert macromolecule such as a glucan polymer is regulated as well as coordinated with the synthesis of other cell wall polymers. Specifically, one would like to know whether this type of self-regulation and cross-regulation is at one or many of the following levels: (1) hexose phosphate intermediates, so that the proper ratio and amount of precursors will flow into the synthetic pathways; (2) nucleotide sugars, in which the amount of one particular nucleotide sugar might regulate the activity of the first enzyme in its own synthesis (Kornfield et al., 1964) and possibly also the synthesis of other sugar nucleotides (Bernstein and Robbins, 1965; Young and Arias, 1967); (3) polymerization, in which a disaccharide such as cellobiose might regulate polymerization of glucose into glucans or acetyl glucosamine monomers into chitin, or where specific inhibitory proteins (Cabib and Keller, 1971) may play a decisive role; (4) secretion, in which a specific glucose-containing lipopolysaccharide might be involved in the active transport of an amino sugar polymer; (5) assembly into cell wall, where a localized enzyme needed for the assembly of one polymer would actually require the presence of a different polymer for activity at the site of cell wall growth or would require the presence of a particular site on the cell wall that was not masked by an inhibitory protein (Yabuki and Fukui, 1970). One could easily imagine more details of regulation at each level, as well as appropriate combinations of levels, but not enough hard facts are available for further speculation.

Some of the other details of cell wall synthesis that are lacking to date pertain to the transport and assembly process. It is not known whether the individual polysaccharide chains assumed to be cytoplasmically synthesized are of the same length as those in the cell wall or whether some part of the polymer is cleaved off during or after secretion. Similarly, one can ask if there are any temporary protecting groups on these polymers which are required for their orientation and binding to lipid or protein within a membrane, and which are then removed. In addition, one can ask whether there are additional chemical moieties which are cleaved off during the assembly process and thus provide the energy for a cross-linking reaction. It should also be pointed out that for most fungal cell walls, it is not known which polymers are covalently attached to the cell wall matrix and which are just held in place by electrostatic and/or hydrogen bonding forces. In addition to this unknown type of permanent bonding, one does not know the extent of the transient forces that order and arrange the newly synthesized components for assembly into the existing cell wall. The information about these types of forces will be critical to the understanding of cell wall elasticity and its mode of growth, and a total understanding of fungal morphogenesis, regardless of the detailed chemistry, will not be complete without this type of physical analysis. In other organisms, the effects of physical forces (e.g., stress lines) on fibril orientation have been reported (Green, 1962, 1965, 1969), but the details for fungal cell walls are still not available. However, one interesting type of analysis which may be of some general applicability to fungal cell walls and other carbohydrate structures deals with the innate properties of different polysaccharide chains, especially when they interact with each other to form fibrils (Rees and Scott, 1969). Up to the present time, the analysis of the morphological mutants has concentrated more on the study of metabolism than on the cell wall changes or morphology changes. Therefore, the emphasis should shift toward gathering more pertinent information, such as the role of assembly enzymes in cell wall formation, the site of synthesis of cell wall components, the three-dimensional structure of the cell wall, and the forces that orient the newly synthesized cell wall components.

This chapter has been purposefully slanted away from the role of gene regulation in morphogenesis. This change in emphasis was done to provide a different look at the subject, since it is so widely accepted and so widely reviewed that gene regulation is of primary importance in this area. To redress the balance of this article a little, one could easily relate the role of the gene to morphogenesis by listing the possible types of genetic information that might be involved in morphogenetic processes. One type of genetic information would be the structural genes for catalytically inert proteins, a type of structural protein. A second type would be the structural genes for the enzymes which catalyze the synthesis of the monomers and polymers. A third type would be the structural genes for the enzymes or proteins which actually are needed to assemble the components into a structure. A fourth type would be the genetic information which regulates the levels of individual gene products from the first three types of genes. A fifth type might be a coordinator type of regulator gene, which would facilitate the simultaneous and orderly expression of metabolically unrelated operons in response to one primary signal. This latter type of gene control could be part of a system of metabolic interlock called into play only at a particular time in the development of an organism. Although all these different types of genes have not been described so far in fungi, it should be apparent that just studying the regulation of the first three types of genes does not indicate how their activation will affect the regulation of many other genes. Perhaps, the time has come to look beyond the details of single operon or gene activation and at those gene interactions which affect the physiology of the cell, and eventually the pattern of organization. It may be that even as we come to understand a particular step in morphogenesis at the biochemical level, we may still be slightly disappointed as Bonner (1960) once implied if the specifics are unique to each organism studied and the generalities few and far between.

VI. Summary

This chapter lists and discusses five questions pertinent to *general* morphogenesis. They are

(1) What role do enzymes play in a shape-determining process?

(2) Can these processes be completely described by a chemical analysis?

(3) Is there strict genetic control of these processes, even at the ultrastructural level?

(4) What is the primary biochemical product of those genes directly involved in these processes?

(5) Does the coordination of these processes occur at the gene level or at the gene product level?

These questions have been discussed in terms of advantages and disadvantages of the genetic, biochemical, cytological, and physical approaches employed in answering these questions. There is also a discussion of the types of experimental manipulations that can be performed on an organism which would be useful in answering these questions. Of the list of useful qualities of an organism, certain genetic properties are considered most important. The useful feature of an organism from a biochemical standpoint is the ability to obtain repeatedly the minimum amount of material necessary for a particular analysis.

These questions are also discussed in a more specific way as relating to the cellular morphogenesis and structural features of certain microorganisms. In fungi and bacteria, the shape-determining structures appear to be the cell walls, and the arguments are presented for the cell wall being the causal factor in fungal morphogenesis. The observations relating cell wall changes and morphological changes in fungi are listed and the role of individual cell wall components are discussed. In addition, the approaches used to answer questions about structure–function relationships for cell walls have been classified into three groups, chemical characterization, mutational approach, and surface structure analysis. The principal specific question that all these approaches must focus on is: How do hyphal branches form?

The use of *Neurospora* as a model system has allowed a great deal of information to be gathered on the biochemical genetics of the development and morphogenesis of this organism. This article classifies the 120 morphological mutants into five general groups according to the affected phase of the life cycle. The genetic nature and the pleiotropic effects of these mutations are also listed. In eleven of these mutants, the primary biochemical deficiency has been detected, and the evidence for these enzyme changes is reviewed. These deficiencies affect either phospholipid or carbohydrate metabolism. The effects of certain genetic interactions, such as reverse mutations, suppressor mutations, and double mutations on particular morphological mutants are also presented. Some of the pleiotropic effects due to a G-6-PD mutation are documented and traced back to a NADPH deficiency. Pleiotropic effects of other morphological mutations are discussed as well as a general scheme for distinguishing the important pleiotropic effects from the trivial ones.

A detailed discussion is presented of the type of information needed for a thorough understanding of the cause and effect relationships in fungal morphogenesis. This discussion centers on our ignorance of the mechanics of cell wall biosynthesis. The numerous morphological mutants already detected indicate the complexity of cell wall assembly and emphasize that one cannot think in terms of a single cause or a single set of relationships in attempting to analyze even the relative simplicity of the formation of a hyphal branch.

Acknowledgments

I would like to acknowledge the helpful discussions of many of my colleagues, in particular, S. Bartnicki-Garcia and W. F. Loomis, Jr. I would also like to thank Stanley Martins for the preparation of the photomicrographs and Gladys Foltz for the careful and patient assistance in the preparation of this manuscript.

References

Abramsky, T., Scott, W. A., and Tatum, E. L. (1971). *Neurospora* morphology and 6-phosphogluconic acid dehydrogenase. *Fed. Proc., Fed. Amer. Soc. Exp. Biol.* **30**, 97 (abstr.).

Ainsworth, G. C., and Sussman, A. S., eds. (1966). "The Fungi," Vol. **2**. Academic Press, New York.

Aronson, J. M. (1965). The cell wall. *In* "The Fungi" (G. C. Ainsworth and A. S. Sussman, eds.), Vol. 1,, pp. 49–76. Academic Press, New York.

Bacon, J. S. D., Farmer, V. C., Jones, D., and Taylor, I. (1969). The glucan components of the cell wall of baker's yeast (*Saccharomyces cerevisiae*) considered in relation to its ultrastructure. *Biochem. J.* **114**, 557–561.

Balassa, G., and Yamamoto, T. (1970). Biochemical genetics of bacterial sporulation. III. Correlation between morphological and biochemical properties of sporulation mutants. *Mol. Gen. Genet.* **108**, 1–22.

Barber, J. T., Srb, A. M., and Steward, F. C. (1969). Proteins, morphology, and genetics in *Neurospora*. *Develop. Biol.* **20**, 105–124.

Barbesgaard, P., and Wagner, S. (1959). Further studies on the biochemical basis of protoperithecia formation in *Neurospora crassa*. *Hereditas* **45**, 564–572.

Barratt, R. W., and Garnjobst, L. (1949). Genetics of a colonial microconidiating mutant strain of *Neurospora crassa*. *Genetics* **34**, 351–369.

Bartnicki-Garcia, S. (1968). Cell wall chemistry, morphogenesis, and taxonomy of fungi. *Annu. Rev. Microbiol.* **22**, 87–108.

Bartnicki-Garcia, S., and Lippman, E. (1969). Fungal morphogenesis: Cell wall construction in *Mucor rouxii*. *Science* **165**, 302–304.

Bartnicki-Garcia, S., and Nickerson, W. J. (1962). Isolation, composition, and structure of cell walls of filamentous and yeast-like forms of *Mucor rouxii*. *Biochim. Biophys. Acta* **58**, 102–119.

Beadle, G. W., and Tatum, E. L. (1941). Genetic control of biochemical reactions in *Neurospora*. *Proc. Nat. Acad. Sci. U.S.* **27**, 499–506.

Bergman, K., Burke, P. V., Cerdá-Olmedo, E., David, C. N., Delbrück, M., Foster, K. W., Goodell, E. W., Heisenberg, M., Meissner, G., Zalokar, M., Dennison, D. S., and Shropshire, W., Jr. (1969). *Phycomyces*. *Bacteriol. Rev.* **33**, 99–157.

Bernstein, R. L., and Robbins, P. W. (1965). Control aspects of uridine 5′-diphosphate glucose and thymidine 5′-diphosphate glucose synthesis by microbial enzymes. *J. Biol. Chem.* **240**, 391–397.

Bienengräber, V., Friemel, H., Brock, J., Kleist, H., and Mücke, D. (1968). Proteine aus *Neurospora crassa*. 3. Agargel- und immunelektrophoretische Analyse von Enzymen und Isoenzymen bei verschiedenen Stämmen von *Neurospora crassa*. *Eur. J. Biochem.* **6**, 356–372.

Bonner, J. T. (1960). The unsolved problem of development. An appraisal of where we stand. *Amer. Sci.* **48**, 514–527.

Brinton, C. C., Jr. (1965). The structure, function, synthesis, and genetic control of bacterial pili and a molecular model for DNA and RNA transport in gram-negative bacteria. *Trans. N.Y. Acad. Sci.* [2] **27**, 1003–1054.

Brody, S. (1969). Mutational interaction at a metabolic branch point in *Neurospora crassa*. *Bacteriol. Proc.* 1969, 122 (abstr.).

Brody, S. (1970a). A key controlling factor in the determination of cellular shape in *Neurospora crassa*. *J. Cell Biol.* **47**, 26a (abstr.).

Brody, S. (1970b). Correlation between reduced nicotinamide adenine dinucleotide phosphate levels and morphological changes in *Neurospora crassa*. *J. Bacteriol.* **101**, 802–807.

Brody, S. (1972). Regulation of pyridine nucleotide levels and ratios in *Neurospora*. *J. Biol Chem.* **247**, 6013–6017.

Brody, S., and Nyc, J. F. (1970). Altered fatty acid distribution in mutants of *Neurospora crassa*. *J. Bacteriol.* **104**, 780–786.

Brody, S., and Tatum, E. L. (1966). The primary biochemical effect of a morphological mutation in *Neurospora crassa*. *Proc. Nat. Acad. Sci. U.S.* **56**, 1290–1297.

Brody, S., and Tatum, E. L. (1967a). On the role of Glucose-6-phosphate dehydrogenase in the morphology of *Neurospora*. *In* "Organizational Biosynthesis" (H. J. Vogel, J. O. Lampen, and V. Bryson, eds.), p. 295. Academic Press, New York.

Brody, S., and Tatum, E. L. (1967b). Phosphoglucomutase mutants and morphological changes in *Neurospora crassa*. *Proc. Nat. Acad. Sci. U.S.* **58**, 923–930.

Brody, S., and Yanofsky, C. (1963). Suppressor gene alteration of protein primary structure. *Proc. Nat. Acad. Sci. U.S.* **50**, 9–16.

Brown, W. C., and Young, F. E. (1970). Dynamic interactions between cell wall polymers, extracellular proteases and autolytic enzymes. *Biochem. Biophys. Res. Commun.* **38**, 564–568.

Butler, G. M. (1961). Growth of hyphal branching systems in *Coprinus disseminatus*. *Ann. Bot. (London)* [N.S.] **25**, 341–352.

Cabib, E., and Bowers, B. (1971). Chitin and yeast budding. Localization of chitin in yeast bud scars. *J. Biol. Chem.* **246**, 152–159.

Cabib, E., and Keller, F. A. (1971). Chitin and yeast budding. Allosteric inhibition of chitin synthetase by a heat-stable protein from yeast. *J. Biol. Chem.* **246**, 167–173.

Cantino, E. C., and Lovett, J. S. (1964). Non-filamentous aquatic fungi: Model systems for biochemical studies of morphological differentiation. *Advan. Morphog.* **3**, 33–93.

Cassady, W. E., and Wagner, R. P. (1971). Separation of mitochondrial membranes of *Neurospora crassa*. I. Localization of L-kynurenine-3-hydroxylase. *J. Cell Biol.* **49**, 536–541.

Chung, P. L., and Trevithick, J. R. (1970). Biochemical and histochemical localization of invertase in *Neurospora crassa* during conidial germination and hyphal growth. *J. Bacteriol.* **102,** 423–429.

Crandall, M. A., and Brock, T. D. (1968). Molecular basis of mating in the yeast *Hansenula wingei. Bacteriol. Rev.* **32,** 139–163.

Crocken, B. J., and Nyc, J. F. (1964). Phospholipid variations in mutant strains of *Neurospora crassa. J. Biol. Chem.* **239,** 1727–1730.

Crocken, B. J., and Tatum, E. L. (1967). Sorbose transport in *Neurospora crassa. Biochim. Biophys. Acta* **135,** 100–105.

Davis, B. D. (1964). Theoretical mechanisms of differentiation. *Medicine (Baltimore)* **43,** 639–649.

Davis, R. H. (1967). Channeling in *Neurospora* metabolism. *In* "Organizational Biosynthesis" (H. J. Vogel, J. O. Lampen, and V. Bryson, eds.), pp. 303–322. Academic Press, New York.

DePamphilis, M. L., and Adler, J. (1971). Attachment of flagellar basal bodies to the cell envelope: Specific attachment to the outer, lipopolysaccharide membrane and the cytoplasmic membrane. *J. Bacteriol.* **105,** 396–407.

DeTerra, N., and Tatum, E. L. (1961). Colonial growth of *Neurospora. Science* **134,** 1066–1068.

DeTerra, N., and Tatum, E. L. (1963). A relationship between cell wall structure and colonial growth in *Neurospora crassa. Amer. J. Bot.* **50,** 669–677.

Dicker, J. W., DeBusk, A. G., and Turian, G. (1970). A nutritional method for the isolation of morphological mutants of *Neurospora crassa. Experientia* **26,** 1154–1155.

Dingle, A. D., and Fulton, C. (1966). Development of the flagellar apparatus of *Naeglaria. J. Cell Biol.* **31,** 43–54.

Doi, R. H., Brown, L. R., Rodgers, G., and Hsu, Y. (1970). *Bacillus subtilis* mutants altered in spore morphology and in RNA polymerase activity. *Proc. Nat. Acad. Sci. U.S.* **66,** 404–410.

Edgar, R. S., and Lielausis, I. (1964). Temperature-sensitive mutants of bacteriophage T4D: Their isolation and genetic characterization. *Genetics* **49,** 649–662.

Ellar, D. H. (1970). The biosynthesis of protective surface structures of prokaryotic and eukaryotic cells. *Symp. Soc. Gen. Microbiol.* **20,** 167–202.

Emerson, S. (1963). Slime, a plasmoid variant of *Neurospora crassa. Genetica* **34,** 162–182.

Endo, A., Kakiki, K., and Misato, T. (1970). Mechanism of action of the antifungal agent polyoxin D. *J. Bacteriol.* **104,** 189–196.

Esser, K., and Kuenen, R. (1967). "Genetics of Fungi" (English translation by E. Steiner). Springer-Verlag, Berlin and New York.

Foster, G. G., and Suzuki, D. T. (1970). T^s mutations in *Drosophila melanogaster.* IV. A mutation affecting eye facet arrangement in a polarized manner. *Proc. Nat. Acad. Sci. U.S.* **67,** 738–745.

Freese, E. B., Fortnagel, P., Schmitt, R., Klofat, W., Chappelle, E., and Picciolo, G. (1968). Biochemical genetics of initial sporulation stages. *In* "Spores" (L. L. Campbell, ed.), Vol. IV, pp. 82–101. Amer. Soc. Microbiol., Bethesda, Md.

Freese, E. B., Cole, R. M., Klofat, W., and Freese, E. (1970). Growth, sporulation, and enzyme defects of glucosamine mutants of *Bacillus subtilis. J. Bacteriol.* **101,** 1046–1062.

Fristrom, J. W. (1965). Development of the morphological mutant *Cryptocephal* of *Drosophila melanogaster. Genetics* **52,** 297–318.

Fristrom, J. W. (1970). The developmental biology of *Drosophila. Annu. Rev. Genet.* **4,** 325–346.

Fuller, R. C., and Tatum, E. L. (1956). Inositol-phospholipid in *Neurospora* and its relationship to morphology. *Amer. J. Bot.* **43**, 361–365.

Garnjobst, L., and Tatum, E. L. (1967). A survey of new morphological mutants in *Neurospora crassa. Genetics* **57**, 579–604.

Garnjobst, L., and Tatum, E. L. (1970). New crisp genes and crisp-modifiers in *Neurospora crassa. Genetics* **66**, 281–290.

Germanier, R. (1959). Untersuchungen uber den Inositol-stoffwechsel bei einem inositol-autotrophen und einem inositol-heterotrophen Stamm von *Neurospora crassa. Arch. Mikrobiol.* **33**, 333–356.

Ghosh, A., Charalampous, F., Sison, Y., and Borer, R. (1960). Metabolic functions of myo-inositol. Cytological and chemical alterations in yeast resulting from inositol deficiency. *J. Biol. Chem.* **235**, 2522–2528.

Ghuysen, J-M. (1968). Use of bacteriolytic enzymes in determination of wall structure and their role in cell metabolism. *Bacteriol. Rev.* **32**, 425–464.

Ghuysen, J-M., Strominger, J. L., and Tipper, D. J. (1968). Bacterial cell walls. *Comp. Bioch.* **26**, Part A, 53–104.

Girbardt, M. (1969). Die Ultrastruktur der Apikalregiön von Pilzhyphen. *Protoplasma* **67**, 413–441.

Glaser, L., and Brown, D. H. (1957). The synthesis of chitin in cell-free extracts of *Neurospora crassa. J. Biol. Chem.* **228**, 729–742.

Gluecksohn-Waelsch, S. (1963). Lethal genes and analysis of differentiation. *Science* **142**, 1269–1276.

Green, P. B. (1962). Mechanism for plant cellular morphogenesis. *Science* **138**, 1404–1405.

Green, P. B. (1965). Pathways of cellular morphogenesis. A diversity in *Nitella. J. Cell Biol.* **27**, 343–363.

Green, P. B. (1969). Cell morphogenesis. *Advan. Plant Physiol.* **20**, 365–394.

Grigg, G. W. (1960a). The control of conidial differentiation in *Neurospora crassa. J. Gen. Microbiol.* **22**, 662–666.

Grigg, G. W. (1960b). Temperature-sensitive genes affecting conidiation in *Neurospora. J. Gen. Microbiol.* **22**, 667–670.

Grove, S. N., Bracker, C. E. (1970). Protoplasmic organization of hyphal tips among fungi: Vesicles and Spitzenkörper. *J. Bacteriol.* **104**, 989–1009.

Gurdon, J. B., and Woodland, H. R. (1968). Cytoplasmic control of nuclear activity in animal development. *Biol. Rev. Cambridge Phil. Soc.* **43**, 233–267.

Hadorn, E. (1961). "Developmental Genetics and Lethal Factors." Wiley, New York.

Halvorson, H., and Swanson, A. (1969). The role of dipicolinic acid in the physiology of spores. *In* "Spores" (L. L. Campbell, ed.), Vol. IV, pp. 121–132. Amer. Soc. Microbiol., Bethesda, Md.

Hanson, R. S., Peterson, J. A., and Youstin, A. A. (1970). Unique biochemical events in bacterial sporulation. *Annu. Rev. Microbiol.* **24**, 53–90.

Hartwell, L. H. (1967). Macromolecule synthesis in temperature-sensitive mutants of yeast. *J. Bacteriol.* **93**, 1662–1670.

Hartwell, L. H., and McLaughlin, C. S. (1968). Temperature-sensitive mutants of yeast exhibiting a rapid inhibition of protein synthesis. *J. Bacteriol.* **96**, 1664–1671.

Hartwell, L. H., Culotti, J., and Reid, B. (1970). Genetic control of the cell-division cycle in yeast. I. Detection of mutants. *Proc. Nat. Acad. Sci. U.S.* **66**, 352–359.

Higgins, M. L., and Shockman, G. D. (1970). Model for cell wall growth of *Streptococcus faecalis. J. Bacteriol.* **101**, 643–648.

Hirota, Y., Ryter, A., and Jacob, F. (1968). Thermosensitive mutants of *E. coli* affected in the processes of DNA syntheses and cellular division. *Cold Spring Harbor Symp. Quant. Biol.* 33, 677–694.

Horowitz, N. H., and Shen, S-C. (1952). *Neurospora* tyrosinase. *J. Biol. Chem.* 197, 513–520.

Hunsley, D., and Burnett, J. H. (1970). The ultrastructural architecture of the walls of some hyphal fungi. *J. Gen. Microbiol.* 62, 203–218.

Inoue, H., and Ishikawa, T. (1970). Macromolecule synthesis and germination of conidia in temperature-sensitive mutants of *Neurospora crassa. Jap. J. Genet.* 45, 357–369.

Kanetsuna, F., and Carbonell, L. M. (1966). Enzymes in glycolysis and the citric acid cycle in the yeast and mycelial forms of *Paracoccidioides brasiliensis. J. Bacteriol.* 92, 1315–1320.

Katz, D., and Rosenberger, R. F. (1970). A mutation in *Aspergillus nidulans* producing hyphal walls which lack chitin. *Biochim. Biophys. Acta* 208, 452–460.

Kellenberger, E. (1968). Polymorphic assemblies of the same major virus protein subunit (Communication No. 7 in "Studies on the morphopoiesis of the head of phage T4"). *In* "Symmetry and Function of Biological Systems at the Macromolecular Level," Nobel Symp. No. 11, pp. 349–366. Wiley, New York.

Kornberg, A., Spudich, J. A., Nelson, D. L., and Deutscher, M. P. (1968). Origin of proteins in sporulation. *Annu. Rev. Biochem.* 37, 51–78.

Kornfeld, S., Kornfeld, R., Neufeld, E. F., and O'Brien, P. J. (1964). The feedback control of sugar nucleotide biosynthesis in liver. *Proc. Nat. Acad. Sci. U.S.* 52, 371–379.

Krulwich, T. A., and Ensign, J. C. (1968). Activity of an autolytic N-acetyl-muramidase during sphere-rod morphogenesis in *Arthrobacter crystallopoietes. J. Bacteriol.* 96, 857–859.

Kushner, D. J. (1969). Self-assembly of biological structures. *Bacteriol. Rev.* 33, 302–345.

Laemmli, U. K., Mölbert, E., Showe, M., and Kellenberger, E. (1970). Form-determining function of the genes required for the assembly of the head of bacteriophage T4. *J. Mol. Biol.* 49, 99–113.

Lauffer, M. A., and Stevens, C. (1968). Structure of the tobacco mosaic virus particle; polymerization of the tobacco mosaic virus protein. *Advan. Virus Res.* 13, 1–63.

Leary, J. V., and Srb, A. M. (1969). Giant spore, a new developmental mutant of *Neurospora crassa. Neurospora Newslett.* 15, 22–23.

Lechner, J. F., and Fuscaldo, K. E. (1969). Analysis of the 6-phosphogluconate dehydrogenase isozymes of *Neurspora crassa. Bacteriol. Proc.* 1969, 120 (abstr.).

Lester, H. E., and Gross, S. R. (1959). Efficient method for selection of auxotrophic mutants of *Neurospora. Science* 129, 572.

Levine, M. (1969). Phage morphogenesis. *Annu. Rev. Genet.* 3, 323–342.

Lie, K. B., and Nyc, J. (1962). The growth of niacin-requiring strains of *Neurospora crassa* with normal and altered phospholipid compositions. *Biochim. Biophys. Acta* 57, 341–347.

Lieberman, M. M., and Markovitz, A. (1970). Depression of guanosine diphosphate-mannose pyrophosphorylase by mutations in two different regulator genes involved in capsular polysaccharide synthesis in *Escherichia coli* K-12. *J. Bacteriol.* 101, 965–972.

Lieberman, M. M., Shaparis, A., and Markovitz, A. (1970). Control of uridine diphosphate-glucose dehydrogenase synthesis and uridine diphosphate-glucuronic acid accumulation by a regulator gene mutation in *Escherichia coli* K-12. *J. Bacteriol.* **101**, 959–964.

Lindberg, A. A., and Hellerqvist, C. G. (1971). Bacteriophage attachment sites, serological specificity, and chemical composition of the lipopolysaccharides of semirough and rough mutants of *Salmonella typhimurium. J. Bacteriol.* **105**, 57–64.

Livingston, L. (1969). Locus specific changes in cell wall composition characteristic of osmotic mutants of *Neurospora crassa. J. Bacteriol.* **99**, 85–90.

Loomis, W. F., Jr. (1969). Temperature-sensitive mutants of *Dictyostelium discoideum. J. Bacteriol.* **99**, 65–69.

Losick, R., and Sonenshein, A. L. (1969). Changes in the template specificity of RNA polymerase during sporulation of *Bacillus subtilis. Nature (London)* **224**, 35–37.

Lysek, G., and Esser, K. (1971). Rhythmic mycelial growth in *Podospora anserina. Arch. Mikrobiol.* **75**, 360–373.

McClure, W. K., Park, D., and Robinson, P. M. (1968). Apical organization in the somatic hyphae of fungi. *J. Gen. Microbiol.* **50**, 177–182.

McMurrough, I., Flores-Carreon, A., and Bartnicki-Garcia, S. (1971). Pathway of chitin synthetase and cellular localization of chitin synthetase in *Mucor rouxii. J. Biol. Chem.* **246**, 3999–4007.

McNelly-Ingle, C. A., and Frost, L. C. (1965). The effect of temperature on the production of perithecia by *Neurospora crassa. J. Gen. Microbiol.* **39**, 33–42.

Mahadevan P. R., and Mahadkar, U. R. (1970). Role of enzymes in growth and morphology of *Neurospora crassa:* Cell-wall-bound enzymes and their possible role in branching. *J. Bacteriol.* **101**, 941–947.

Mahadevan, P. R., and Menon, C. P. S. (1968). Laminarinase of *Neurospora crassa.* Part I. Enzyme activity associated with conidia and conidial wall. *Indian J. Biochem.* **5**, 6–8.

Mahadevan, P. R., and Tatum, E. L. (1965). Relationship of the major constituents of the *Neurospora crassa* cell wall to wild-type and colonial morphology. *J. Bacteriol.* **90**, 1073–1081.

Mahadevan, P. R., and Tatum, E. L. (1967). Localization of structural polymers in the cell wall of *Neurospora crassa. J. Cell Biol.* **35**, 295–302.

Manocha, M. S., and Colvin, J. R. (1967). Structure and composition of the cell wall of *Neurospora crassa. J. Bacteriol.* **94**, 202–212.

Marchant, R., and Smith, D. G. (1968). A serological investigation of hyphal growth in *Fusarium culmorum. Arch. Mikrobiol.* **63**, 85–94.

Markovitz, A. (1964). Regulatory mechanisms for synthesis of capsular polysaccharide in mucoid mutants of *Escherichia coli* K-12. *Proc. Nat. Acad. Sci. U.S.* **51**, 239–246.

Matile, P. (1966). Inositol deficiency resulting in death: An explanation of its occurrence in *Neurospora crassa. Science* **151**, 86–88.

Mishra, N. C., and Tatum, E. L. (1970). Phosphoglucomutase mutants of *Neurospora sitophila* and their relation to morphology. *Proc. Nat. Acad. Sci. U.S.* **66**, 638–645.

Mishra, N. C., and Threlkeld, S. F. H. (1967). Variation of the expression of the ragged mutant in *Neurospora. Genetics* **55**, 113–121.

Morton, A. G. (1967). Morphogenesis in fungi. *Sci. Progr. (London)* **55**, 597–611.

Murayama, T. (1969). Biochemical and genetical studies on morphological change in a phosphohexoisomerase mutant of *Neurospora. Jap. J. Genet.* **44**, 399. (Abstracts of papers presented at the 41st Annual Meeting of the Genetic Society of Japan.)

Neidhardt, F. C. (1966). Roles of amino acid activating enzymes in cellular physiology. *Bacteriol. Rev.* **30**, 701–719.

Nelson, D. L., Spudich, J. A., Bonsen, P. P. M., Bertsch, L. L., and Kornberg, A. (1969). Biochemical studies of bacterial sporulation and germination. XVI. Small molecules in spores. *In* "Spores" (L. L. Campbell, ed.), Vol. IV, pp. 59–71. Amer. Soc. Microbiol., Bethesda, Md.

Nickerson, W. J. (1964). Composition and function of microbial cell walls. *In* "Cellular Membranes in Development," (M. Locke, ed.), pp. 281–297. Academic Press, New York.

Nickerson, W. J., Falcone, G., and Kessler, G. (1961). Polysaccharide protein complexes of yeast cell walls *In* "Macromolecular Complexes" (M. V. Edds, Jr., ed.), pp. 205–228. Ronald Press, New York.

Niederpruem, D. J., and Wessels, J. G. H. (1969). Cytodifferentiation and morphogenesis in *Schizophyllum commune. Bacteriol. Rev.* **33**, 505–535.

Novak, D. R., and Srb, A. M. (1971a). A developmental mutation that affects ascospore delimitation in *Neurospora tetrasperma. Can. J. Genet. Cytol.* **13**, 1–8.

Novak, D. R., and Srb, A. M. (1971b). Genetic alterations of ascus development in *N. tetrasperma. Genetics* **67**, 201–208.

Nyc, J., and Brody, S. (1971). Effects of mutations and growth conditions on lipid synthesis in *Neurospora. J. Bacteriol.* **108**, 1310–1317.

Osborn, M. J. (1969). Structure and biosynthesis of the bacterial cell wall. *Annu. Rev. Biochem.* **38**, 501–538.

Ojha, M. N., Velmelage, R., and Turian, G. (1969). Melonate metabolism, stimulation of conidiation and succinate dehydrogenase activity in *Neurospora crassa. Physiol. Plant.* **22**, 819–826.

Park, D., and Robinson, P. M. (1966). Internal pressure of hyphal tips of fungi, and it significance in morphogenesis. *Ann. Bot. (London)* [N.S.] **30**, 425–439.

Pearce, S. M., and Fitz-James, P. C. (1971). Sporulation of a cortexless mutant of a variant of *Bacillus cereus. J. Bacteriol.* **105**, 339–348.

Peduzzi, R., and Turian, G. (1969). Etude immunochimique et électrophorétique de la conidiation de *Neurospora crassa. Experientia* **25**, 1178–1180.

Perkins, D. D., Newmeyer, D., Taylor, C. W., and Bennett, D. C. (1969). New markers and map sequences in *Neurospora crassa*, with a description of mapping by duplication coverage, and of multiple translocation stocks for testing linkage. *Genetica* **40**, 247–278.

Pina, E., and Tatum, E.L. (1967). Inositol biosynthesis in *Neurospora crassa. Biochim. Biophys. Acta.* **136**, 265–271.

Pincheira, G., and Srb, A. M. (1969a). Cytology and genetics of two abnormal ascus mutants of *Neurospora. Can. J. Genet. Cytol.* **11**, 281–286.

Pincheira, G., and Srb, A. M. (1969b). Genetic variation in the orientation of nuclear spindles during the development of asci in *Neurospora. Amer. J. Bot.* **56**, 846–852.

Pitel, D. W., and Gilvarg, C. (1971). Timing of mucopeptide and phospholipid synthesis in sporulating *Bacillus megatherium. J. Biol. Chem.* **246**, 3720–3724.

Power, D. M., and Challinor, S. W. (1969). The effects of inositol-deficiency on the chemical composition of the yeast cell wall. *J. Gen. Microbiol.* **55**, 169–176.

Rees, D. A., and Scott, W. A. (1969). Conformational analysis of polysaccharides:

Stereochemical significance of different linkage positions in β-linked polysaccharides. *Chem. Commun.* pp. 1073–1038.

Ritossa, F. M., Atwood, K. C., and Spiegelman, S. (1966). A molecular explanation of the bobbed mutants of *Drosophila* as partial deficiencies of "ribosomal" DNA. *Genetics* **54**, 819–834.

Rizki, M. T. M. (1961). The influence of glucosamine-hydrochloride on cellular adhesiveness in *Drosophila melanogaster. Exp. Cell Res.* **24**, 111–119.

Rizvi, S. R. H., and Robertson, N. F. (1965). Apical disintegration of hyphae of *Neurospora crassa* as a response to 1-sorbose. *Trans. Brit. Mycol. Soc.* **48**, 469–477.

Roberts, R., Abelson, R., Cowie, D., Bolton, E., and Butten, R. (1955). Studies of biosynthesis in *E. coli. Carnegie Inst. Wash. Publ.* **607.**

Robertson, N. F. (1968). The growth in fungi. *Annu. Rev. Phytopathol.* **6,** 115–136.

Rogers, H. J. (1970). Bacterial growth and the cell envelope. *Bacteriol. Rev.* **34,** 194–214.

Rothschild, H., and Suskind, S. R. (1966). Protoperithecia in *Neurospora crassa:* Technique for studying their development. *Science* **154,** 1356–1357.

Sargent, M. L., Briggs, W. R., and Woodward, D. O. (1966). Circadian nature of a rhythm expressed by an invertaseless strain of *Neurospora crassa. Plant Physiol.* **41,** 1343–1349.

Scarborough, G. A., and Nyc, J. F. (1967). Properties of a phosphatidyl-monomethyl-ethanolamine *N*-methyltransferase from *Neurospora crassa. Biochim. Biophys. Acta* **146,** 111–119.

Schmidt, J. M., and Stanier, R. Y. (1966). The development of cellular stalks in bacteria. *J. Cell Biol.* **28,** 432–436.

Schwarz, U. Asmus, A., and Frank, H. (1969). Autolytic enzymes and cell division of *Escherichia coli. J. Mol. Biol.* **41,** 419–429.

Scott, W. A., and Tatum, E. L. (1970). Glucose-6-phosphate dehydrogenase and *Neurospora* morphology. *Proc. Nat. Acad. Sci. U.S.* **66,** 515–522.

Sheng, T. C., and Ryan, F. J. (1948). Mutations involving the production of conidia and the requirement for leucine in a mutant of *Neurospora. Genetics* **33,** 221–227.

Siegel, R. W., Matsuyama, S. S., and Urey, J. C. (1968). Induced macroconidia formation in *Neurospora crassa. Experientia* **24,** 1179–1181.

Srb, A. M., and Basl, M. (1969). The isolation of mutants affecting ascus development in *Neurospora crassa* and their analysis by a zygote complementation test. *Genet. Res.* **13,** 303–311.

Stern, C. (1969). Gene expression in genetic mosaics. *Genetics* **61,** Suppl., 199–211.

Sussman, A. S., von Boventer-Heidenhain, B., and Lowry, R. J. (1957). Physiology of the cell surface of *Neurospora* ascospores. IV. The functions of surface binding sites. *Plant Physiol.* **32,** 586–590.

Sussman, A. S., Lowry, R. J., and Durkee, T. (1964). Morphology and genetics of a periodic colonial mutant of *Neurospora crassa. Amer. J. Bot.* **51,** 243–252.

Sussman, M. (1967). Evidence for temporal and quantitative control of genetic transcription and translation during slime mold development. *Fed. Proc., Fed. Amer. Soc. Exp. Biol.* **26,** 77–83.

Suzuki, D. T. (1970). T^s mutations in *Drosophila melanogaster. Science* **170,** 695–706.

Szulmajster, J., Bonamy, C., and Laporte, J. (1970). Isolation and properties of a temperature-sensitive sporulation mutant of *Bacillus subtilis. J. Bacteriol.* **101,** 1027–1037.

154 STUART BRODY

Tatum, E. L., Barratt, R. W., and Cutter, V. M. (1949). Chemical induction of colonial paramorphs in *Neurospora* and *Syncephalastrum*. *Science* 109, 509–511.

Tee, T. S., and Choke, H. C. (1970). A gene controlling the early development of protoperithecium in *Neurospora crassa*. *Mol. Gen. Genet.* 107, 158–161.

Tipper, D. J., and Pratt, I. (1970). Cell wall polymers of *Bacillus sphaericus* 9602. II. Synthesis of the first enzyme unique to cortex synthesis during sporulation. *J. Bacteriol.* 103, 305–317.

Tkacz, J. S., Cybulska, E. B., and Lampen, J. O. (1971). Specific staining of wall mannan in yeast cells with fluorescein-conjugated concanavalin A. *J. Bacteriol.* 105, 1–5.

Tomasz, A. (1968). Biological consequences of the replacement of choline by ethanolamine in the cell wall of pneumococcus: Chain formation, loss of trans-formability, and loss of autolysis. *Proc. Nat. Acad. Sci. U.S.* 59, 86–93.

Trevithick, J. R., and Metzenberg, R. L. (1966). Molecular sieving by *Neurospora* cell walls during secretion of invertase isozymes. *J. Bacteriol.* 92, 1010–1015.

Trinci, A. P. J. (1969). A kinetic study of the growth of *Aspergillus nidulans* and other fungi. *J. Gen. Microbiol.* 57, 11–24.

Troy, F. A., and Koffler H. (1969). The chemistry and molecular architecture of the cell walls of *Penicillium chrysogenum*. *J. Biol. Chem.* 244, 5563–5576.

Turian, G. (1969). "Differenciation fongique." Masson et Cie, Paris.

Wang, M. C., and Bartnicki-Garcia, S. (1966). Biosynthesis of β-1,3- and β-1-6-linked glucan by *Phytophthora cinnamomi* hyphal walls. *Biochem. Biophys. Res. Commun.* 24, 832–837.

Weston, C. R. (1965). Morphogenesis in fungi: A study of hyphal branching. Ph.D. Dissertation, Princeton University, Princeton, New Jersey, 1965. (University Microfilms, Ann Arbor, Michigan.)

Wilson, J. F. (1961). Micrurgical techniques for *Neurospora*. *Amer. J. Bot.* 48, 46–51.

Wilson, R. W., and Niederpruem, D. J. (1967a). Control of β-glucosidases in *Schizophyllum commune*. *Can. J. Microbiol.* 13, 1009–1020.

Wilson, R. W., and Niederpruem, D. J. (1967b). Cellobiose as a paramorphogen in *Schizophyllum commune*. *Can. J. Microbiol.* 13, 1663–1670.

Wood, W. B., Edgar, R. S., King, J., Lielausis, I., and Henninger, M. (1968). Bacteriophage assembly. *Fed. Proc. Fed. Amer. Soc. Exp. Biol.* 27, 1160–1166.

Yabuki, M., and Fukui, S. (1970). Presence of binding site for α-amylase and of masking protein for this site on mycelial cell wall of *Aspergillus oryzae*. *J. Bacteriol.* 104, 138–144.

Yanagida, M., Boy de la Tour, E., Alff-Steinberger, C., and Kellenberger, E. (1970). Studies on the morphopoiesis of the head of bacteriophage T-even. VIII. Multilayered polyheads. *J. Mol. Biol.* 50, 35–58.

Young, F. E., and Arias, L. (1967). Biosynthesis of the N-acyl-galactosamine in cell walls of *Bacillus subtilis*. *J. Bacteriol.* 94, 1783–1784.

Zalokar, M. (1959). Enzyme activity and cell differentiation in *Neurospora*. *Amer. J. Bot.* 46, 555–559.

6

Colony Differentiation in Green Algae

GARY KOCHERT

I. Introduction

The green algae of the family Volvocaceae have long been regarded as a classic example of a group of organisms constituting an evolutionary pathway. The organisms in this family are motile, flagellated colonies in which the individual cells are embedded in a transparent gelatinous sheath. The more primitive genera consist of small numbers of cells which are morphologically and functionally similar. In these primitive genera, all cells function in both asexual and sexual reproduction. In the more advanced members, larger numbers of cells are present and the cells are differentiated into two types: somatic and reproductive. In colonies of the genus *Volvox*, which is considered to be the most highly evolved member of this series, there are typically hundreds or even thousands of somatic cells and only a few reproductive cells. The somatic cells appear to be terminally differentiated and normally die without further division after the colonies undergo either sexual or asexual reproduction.

Volvox is an organism, then, which possesses only two cell types. These cell types are very different, both in morphology and function, and the organism represents cell differentiation in one of its simplest forms. This fact coupled with the ease of growing large populations of genetically identical organisms under controlled conditions has made *Volvox* an increasingly important object for studies of cell differentiation.

155

There are numerous species of *Volvox* and these present a fascinating variety of developmental patterns. Details of differentiation in the various species have been recently reviewed (Starr, 1968, 1970; McCracken, 1970). The purpose of the present chapter is to discuss the evolution of developmental patterns in the Volvocaceae and aspects of differentiation in one species of *Volvox* in somewhat more detail.

II. *Chlamydomonas* and *Gonium*

The ancestor of the Volvococean colonial forms is believed to have been a unicellular form similar to the extant genus *Chlamydomonas*. This organism is anteriorly biflagellate and has a single cup-shaped chloroplast containing a conspicuous orange eyespot. *Chlamydomonas* reproduces asexually by zoospore formation. During this process the protoplast of the parent cell cleaves into a number of small zoospores each of which is essentially a miniature replica of the parent cell. These are at first retained inside the parent cell wall where they swim more or less freely, but are eventually released by dissolution of the wall (Fig. 1).

Sexual reproduction in *Chlamydomonas* is usually triggered by depletion of nitrogen in the culture medium (Sager and Granick, 1954). During sexual reproduction cells differentiate to form gametes. The gametes differ from vegetative cells in the possession on their flagella of sexual substances termed gamones (Weise, 1969). In addition, some species possess a fertilization tubule which is the site where cell fusion is initiated (Friedmann *et al.*, 1968). The gametes fuse in pairs to form resistant zygospores. Most species of *Chlamydomonas* exhibit isogamous sexual reproduction; the members of a fusing pair of gametes do not differ in size or form (Fig. 2).

Gonium, the simplest colonial member of the volvocine line, has cells embedded in a gelatinous sheath material and arranged in a flat plate (Fig. 1). Each cell of the colony is identical in overall appearance to a cell of *Chlamydomonas*, and all cells in the colony are morphologically identical; no cell specialization has taken place. *Gonium* reproduces asexually by daughter colony formation, as do all the colonial Volvocaceae. In this process each parent cell of the colony cleaves into smaller cells which remain associated with each other and are released from the parent together as a colony. Normally all the cells of the colony form daughter colonies at about the same time. During this process, the entire contents of the parent cell are used to produce a single daughter colony. Daughter colonies are formed by a series of synchronous cell divisions. Thus,

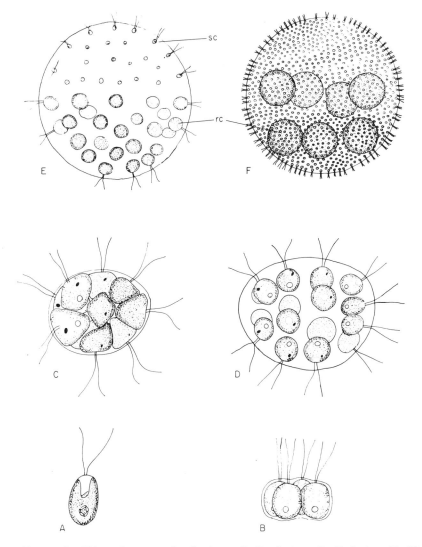

Fig. 1. A, *Chlamydomonas;* B, *Gonium;* C, *Pandorina;* D, *Eudorina;* E, *Pleodorina;* F, *Volvox;* sc = somatic cell; rc = reproductive cell.

colonies normally contain 2^n cells. Just how many cells the colony contains is a reflection of how large the parent cell was which produced it. The larger the parent cell the greater the number of cells in the resultant daughter colony; cell divisions appear to take place until a certain minimum cell size is reached. Divisions then cease and the newly formed colony is eventually released by dissolution of the parent cell

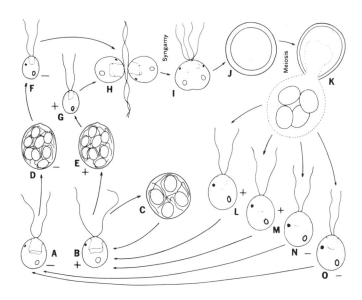

FIG. 2. Life cycle of *Chlamydomonas*. A, B, Vegetative cells; C, zoospore produc-tion; D, E, gamete formation; F, G, gametes; H, I, gamete fusion; J, zygospore; K, zygospore germination; L–O, meiospores (+ and − indicate mating types). (From Scagel *et al.,* "An Evolutionary Survey of the Plant Kingdom," reproduced by permission of Wadsworth Publishing Co.)

wall. After the daughter colony has been released it grows in size by enlargement of the individual cells and by secretion of the sheath mate-rial between the cells. The latter process effectively moves the cells fur-ther apart. No increase in cell number takes place after the colony is formed.

Gonium reproduces sexually in a manner very similar to *Chlamy-domonas*. Individual cells escape from the colony and fuse in pairs to form zygotes. Just as in *Chlamydomonas* the gametes are isogamous.

Thus, in most aspects *Gonium* is a very simple colony representing little more than a grouping of *Chlamydomonas*. All cells of the colony act in an independent and identical fashion during asexual and sexual reproduction.

III. *Pandorina, Eudorina,* and *Pleodorina*

The next step in colonial evolution is represented by the genus *Pan-dorina*. Colonies of this organism are spherical to oval and typically contain 8–32 cells. The cells are embedded in sheath material and arranged in a single layer on the colony surface (Fig. 1). Even when the colony is spherical a definite polarity is present. There is an anterior pole which is always directed toward the front as the colony swims. During forward motion the colony rotates around its longitudinal axis.

It is in the genus *Pandorina* that the first signs of cell differentiation within a colony are seen. This differentiation involves the relative size of the eyespot (stigma). Cells near the front of *Pandorina* colonies have large and conspicuous eyespots, while those nearer the posterior pole of the colony have relatively smaller eyespots. This organelle, which in green algae consists of a number of pigment granules arranged in the chloroplast between the photosynthetic lamellae, is thought to be involved in phototactic responses.

All cells of *Pandorina* colonies function in daughter colony formation. During this process the cells divide regularly in two planes to form a flat sheet of cells termed a plakea (Fig. 3). The plakea becomes curved into a cup shape during its formation, presumably because of the space limitation imposed by the parent cell wall. The plakea always curves in such a way that the future flagellar ends of the cells are pointed inward toward the hollow interior of the plakea. This state of affairs is corrected by the process called inversion which is characteristic of this group of organisms. During inversion the plakea undergoes a series

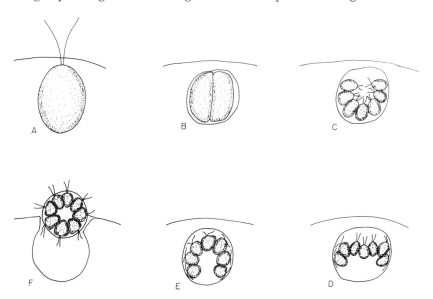

Fig. 3. Stages in asexual reproduction of a colonial green alga. A, Parent cell in colony sheath; B, 2-celled stage; C, curved plakea stage; D, inversion in process; E, inversion nearly complete; F, escape from the parent colony.

of morphogenetic movements which result in its turning itself inside out so that the flagellar ends of the cells are now directed outward. Inversion is accomplished in the colonial Volvococeae without relative cell movement; only bending and folding of the plakea is involved. In this respect the process bears only a superficial resemblance to gastrulation as seen in animals.

During sexual reproduction in *Pandorina,* individual cells escape from the colony and function directly as isogamous gametes. Apparently all the cells of the colony have equal potential for sexual reproduction.

The spherical colony shape adopted by *Pandorina* probably has significance in determining the polarity of daughter colonies. A cell on the periphery of a spherical colony is exposed to environmental gradients of such components as light and nutrient concentration. These concentration gradients could be the basis for the establishment of such features of colony polarity as the range of eyespot size discussed above. This would be "programmed" into the cell before it divides by the environmental gradients imposed on it because of its location in the colony. Evidence for such preprogramming will be discussed below in the case where it has been documented by experimentation.

Individuals of the genus *Eudorina* contain 16–64 cells in a spherical colony. (Fig. 1). The cell number is greater than in *Gonium* because the cells enlarge before cleavage to form daughter colonies; more material is thus available for the formation of cells. Also the minimum size which cells reach before cleavage ends has been reduced in *Eudorina* and other members of the Volvocaceae thereby increasing the possible cell number.

In *Eudorina,* cell differentiation in the colony is more pronounced than in *Pandorina.* The same sort of anterior–posterior eyespot size gradient is observed, but in addition one species shows a specialization of cell function. In this species the anterior four cells of the colony are reduced in size (because they failed to enlarge as much as the other cells). When the colony undergoes daughter colony formation this group of four cells is much delayed in division and often do not form daughter colonies at all. Thus, in this species the first beginnings of a differentiation into somatic and reproductive cell lines are observed. This tendency, which is more characteristic of animals than plants, is highly developed in the more advanced members of the family Volvocaceae

Sexual reproduction in *Eudorina* is also of a relatively advanced type. Cells producing male gametes form in a cleavage process similar to daughter colony formation a bundle of sperm which is apparently a highly reduced colony. This assemblage of male gametes is released as a unit and swims about until encountering a female colony. Upon encountering a female colony the sperm bundle dissociates and individual sperm penetrate the colony to fuse with the female gametes. Female gametes are morphologically similar to an ordinary somatic cell.

The tendency toward greater cell numbers and toward differentiation into somatic and reproductive cell lines, which can be recognized in *Eudorina*, is amplified in *Pleodorina* (Fig. 1). In *Eudorina* the distinction between somatic and reproductive cells is not great, but in *Pleodorina* there is always a definite distinction between somatic and reproductive cells. This organism contains 32–128 cells arranged in a spherical colony generally similar to *Eudorina* in morphology. When daughter colonies are newly released, all cells of the colony are similar in size and overall appearance. As the colony grows in size the cells in the posterior half of the colony enlarge to a much greater degree than those in the anterior half and may lose their flagella. Only these larger cells (which have been named gonidia) function in daughter colony formation. The smaller cells in the anterior portion of the colony are purely somatic in function and eventually die without further divisions. Thus an elemental division of labor is achieved. Sexual reproduction is, in general, similar to that described for *Eudorina*.

IV. Volvox

The apex of the volvocine line is reached with the genus *Volvox*. It is in this genus that all the evolutionary trends observable in the other genera reach their peak. The smaller species of *Volvox* contain about 1000–3000 cells. Larger species may have up to 50,000 cells. There is always a differentiation of the cells of the colony into gonidia and somatic cells. Many more somatic cells than gonidia are present in a colony. Only the gonidia function in daughter colony formation. The somatic cells appear to be terminally differentiated and normally do not divide again after the young colony has been released from the parent. In order to produce the large number of cells found in the *Volvox* colony, the gonidia have had to enlarge to provide sufficient starting material. Different species of *Volvox* have solved this problem in slightly different ways. In one group of species (which form a separate section of the genus) the gonidia enlarge greatly before divisions begin. The ensuing divisions to form a daughter colony occur very rapidly without apparent growth occurring. This form of cleavage without growth is rare in plants and is more similar to that seen in early stages of frog and sea urchin embryogenesis than that of higher plants. In the other group of *Volvox* species, divisions begin while the gonidium is still quite small. Growth then proceeds while divisions are taking place, in a manner more typical for higher plants. In both cases, however, all the cells to be present in the mature colony are formed before the colony is released from the parent.

There is also a considerable amount of variation among *Volvox* species in the time during daughter colony formation when the somatic cells become visibly differentiated from the reproductive cells. Some species differentiate the two cell types very early. In *Volvox carteri* the gonidia are differentiated from the somatic cells by unequal divisions beginning at the division from 16 to 32 cells. In this species gonidia are formed in a very characteristic pattern during daughter colony formation. This pattern can be recognized in adult colonies. It consists of alternating tiers of four gonidia each (Fig. 4), usually with two or three tiers of four gonidia in each colony. Since these gonidia are differentiated at such an early stage in daughter colony formation, factors controlling their characteristic localization must be present during the very early stages of cleavage or in the uncleaved gonidium. The fact that such a pattern determining the localization of gonidia in the developing daughter colony is present before cleavage begins can be demonstrated by ultraviolet irradiation experiments, the rationale for which can be stated as follows: If a cytoplasmic morphogenetic substance controlling gonidium localization during daughter colony formation is present in the undivided, mature gonidium, it should be possible to inactivate this

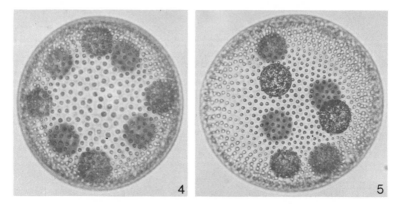

FIG. 4. Anterior polar view of a colony of *Volvox*. The large reproductive cells (gonidia) in the interior portion of the colony are arranged in two tiers of four each. The small somatic cells on the surface of the colony are very similar to *Chlamydomonas* in morphology.

FIG. 5. Side view of a daughter colony developed from an ultraviolet-irradiated gonidium. The anterior (left) tier of four gonidia is normal, but the posterior tier is missing one gonidium.

substance by ultraviolet irradiation. Since ultraviolet is weakly penetrating it should be possible to destroy substances contained in the cytoplasm of a large cell without damaging the nucleus of the cell.

If mature, but undivided gonidia of *Volvox carteri* are treated with short doses of ultraviolet irradiation and then allowed to cleave to form daughter colonies, many of the resultant colonies lack one or more gonidia (Fig. 5).

These colonies are normal in every other respect (Kochert and Yates, 1970). The somatic cells which occupy positions directly over the missing gonidia are not noticeably changed. Thus, the effect is not attributable to general destruction of cytoplasm in the irradiated area of the gonidium since the irradiated cytoplasm still forms seemingly normal somatic cells. Neither is it explainable by nuclear mutation since all the remaining gonidia in the colony (which have nuclei which are descendents of the single nucleus of the original irradiated gonidium) produce wholly normal daughter colonies with no missing gonidia in the next generation. Therefore, it appears that there is present in undivided gonidia a morphogenetic substance which determines the location of the reproductive cells in the daughter colony produced by that gonidium. Presumably this substance is parceled out to cells during cleavage. Nuclei of those cells which receive the substance are "determined" by it to function as gonidial nuclei. Nuclei in cells not receiving the substance (or receiving some other substance) behave as nuclei of somatic cells. The localization of this morphogenetic substance is apparently set up during the long period of gonidial enlargement that takes place before division begins.

The developmental fate of cells of the adult colony is, then, traceable to particular regions of the gonidial cytoplasm. This sort of a nuclear–cytoplasmic interaction is very common in animal development, where cytoplasmic patterns set up during oogenesis determine the developmental fate of specific areas of the embryo (for review, see Gurdon and Woodland, 1968). This has been clearly shown in amphibian germ cell development. In the frog a special staining area of cytoplasm is distributed during embryogenesis in such a way that it ends up in the presumptive gonads of the animal. This material has been shown to be necessary for proper germ cell development. If it is destroyed by ultraviolet irradiation the animal develops normally, but is sterile (Smith, 1966). This sort of nuclear–cytoplasmic interaction is very rare in plant development, however. *Volvox* is an exception to the general rule in plants that no specific germ plasm or germ line can be implicated in reproductive cell formation. In plants, cells which form somatic structures can be induced by the appropriate stiumli to become cells of reproductive structures and even the actual germ cells. Thus *Volvox*, a green plant, has evolved a developmental mechanism more in common with animals than plants.

Volvox carteri, then, apparently preprograms at least two aspects of its colony morphology into the gonidium before it begins to divide: (1) the total number of cells to be present in the colony and (2) the number and distribution of the reproductive cells. It also seems possible that the eyespot size gradient observed in adult colonies is also programmed into the gonidium before cleavage. The underlying mechanism for this could be a gradient of some morphogenetic substance set up in the gonidium in response to an environmental gradient. Developing gonidia are, by virtue of their localization in the parent colony, automatically subject to a light gradient. During gonidial development the anterior end of the gonidium (which becomes the posterior end of the daughter colony) is exposed to relatively more light. Thus, merely by being placed in the colony in such a way as to ensure their exposure to an environmental gradient, gonidia can be programmed to produce a daughter colony with a polarized distribution of eyespots.

Eyespots appear to be important in controlling flagellar motion in cells which contain them, as many studies with unicellular algae have indicated. In normal *Volvox* colonies eyespots are not only present in an anterior–posterior gradient, but also are characteristically positioned in each cell relative to the anterior–posterior axis of the cell. The eyespot is always placed in the lobe of the chloroplast on the side of the cell away from the anterior pole of the colony. These observations would appear to explain the seeming coordination in flagellar beat among the hundreds of cells of a *Volvox* colony which results in the orderly swimming motion of the colony. The cells are simply placed in the colony in such a way and with their eyespots in such a position so that the net result of the cells reacting to an external stimulus (in this case light) is individual patterns of flagellar motion which collectively move the colony in a seemingly coordinated fashion. This situation is rather like the motion of a Roman galley propelled by hundreds of rowers. The rowers need not communicate one with another; each individually rows in cadence with an audible signal. The net result is coordinated motion of the ship without communication between the individual rowers.

Thus far we have discussed only asexual colonies of *Volvox*, but male and female colonies are also formed by many species. *Volvox carteri* forms dwarf male colonies which contain about 50 reproductive cells instead of the 8–12 characteristic of asexual colonies and a relatively small number (several hundred) of somatic cells. The colonies have fewer somatic cells (and hence are dwarf) because a greater proportion of the parent cell material is used to form reproductive cells. Female colonies contain 16–20 eggs in addition to somatic cells. In both male and female colonies the somatic cells are very similar to those found in asexual colonies. These sexual colonies are formed by cleavage of a gonidium as are asexual colonies. The cleavage process is generally similar in each case, but the egg and sperm initial cells become differentiated from the somatic cells at a slightly later stage in cleavage than do the gonidia. Also, the eggs and sperm initials are not formed in regular tiers as are the gonidia in asexual colonies, but are scattered through the posterior portion of the colony. Whether a gonidium will cleave to form an asexual, male, or female colony is also programmed into the gonidium before it begins to cleave. This is most clearly demonstrated in the case of female colonies.

In *Volvox carteri* female colonies normally arise only in response to a substance released into the culture medium by male colonies. This female-inducing substance appears to be a basic protein. Its activity is destroyed by treatment with the proteolytic enzyme Pronase, but is unaffected by RNase or DNase. On the basis of gel chromatography its molecular weight appears to be in the range of 75–100,000. The inducer exerts its effect on young gonidia in the early stages of their long period of enlargement prior to cleavage. Gonidia treated with the female inducer at this early stage cleave to form female rather than asexual colonies. Treatment of gonidia by the inducer must take place at an early stage, while the gonidia are small, to induce them to form female colonies. By the time the gonidia are enlarged prior to cleavage they can no longer be induced to form female rather than asexual colonies. This is consistent with the idea that mature gonidia exhibit cytoplasmic properties which determine the type of colony they will produce on cleaving. Gonidia normally set up a cytoplasmic pattern which leads to the formation of an asexual colony. If treated with the female inducer at an early stage in enlargement, however, the cytoplasmic pattern is changed in such a way that a female rather than an asexual colony is ultimately formed.

Thus, the number of cells in a colony of *Volvox carteri*, as well as the type and location of the reproductive cells, is programmed into the gonidium before it begins to cleave. It also seems probable, although there are no confirmatory experimental data, that the distribution and size of the eyespot among the cells of the colony are programmed. These features include essentially all there is to the morphology of a *Volvox carteri* colony. Colony development in this species appears, therefore, to be completely determined or preprogrammed.

The classic cases of cytoplasmic preprogramming of development have been described in animals such as sea urchins and frogs (Smith, 1966;

Gurdon and Woodland, 1968; Davidson, 1969). In cases such as these, morphogenetic substances present in the egg cytoplasm are important in early embryonic development. In the more advanced organisms, however, other developmental mechanisms come into play in later stages of development. Some of the most important of these involve interactions between the cells and tissues of the developing organism. The phenomona collectively described as embryonic induction are examples of such interactions (Saxén and Toivonen, 1962).

In *Volvox* this switch during development to a different type of developmental mechanism does not occur. Colonial morphology is preprogrammed in the gonidium. After this program has been followed and the colony set up, the component cells behave as essentially independent individuals, not as well-integrated, interacting units of an organism. The very structure of the colony would probably preclude cell–cell interactions since the individual cells are essentially isolated from one another by the noncellular colonial matrix surrounding each cell. Thus, the intimate cell–cell contacts characteristic of higher plants and animals are not possible in the later developmental stages of *Volvox*.

Volvox is considered to be the end member of the volvocine evolutionary line of the green algae. As such it appears to represent an evolutionary dead end. The evolutionary trends exhibited in this group include a tendency toward greater cell numbers and a division of labor involving the formation of separate reproductive and somatic cells. These goals have been accomplished in *Volvox carteri* by formation of a very large gonidium (to provide enough material for the formation of a large number of cells), and by preprogramming the basic pattern of the colony into the gonidium. This organism represents an evolutionary dead end because to construct a larger and more complex organism it would be necessary to provide for continued increase in cell number and to evolve developmental mechanisms involving cell–cell interactions within the colony. *Volvox* has failed to evolve these and therefore is classed as a "primitive" organism.

References

Davidson, E. (1969). "Gene Activity in Early Development." Academic Press, New York.

Friedmann, I., Colwin, A. L., and Colwin, L. (1968). *J. Cell Sci.* **3**, 115.

Gurdon, J. B., and Woodland, H. R. (1968). *Biol. Rev.* **43**, 233.

Kochert, G., and Yates, I. (1970). *Develop. Biol.* **23**, 128.

McCracken, M. (1970). *Carol. Tips* **33**, 37.

Sager, R., and Granick, S. (1954). *J. Gen. Physiol.* **37**, 729–742.

Saxén, L., and Toivonen, S. (1962). "Primary Embryonic Induction." Academic Press, New York.

Smith, L. D. (1966). *Develop. Biol.* **14,** 330.

Starr, R. C. (1968). *Proc. Nat. Acad. Sci. U.S.* **59,** 1082.

Starr, R. C. (1970). *Develop. Biol. Suppl.* **4,** 59.

Weise, L. (1969). *In* "Fertilization: Comparative Morphology, Biochemistry and Immunology" (C. B. Metz and A. Monroy, eds.), Vol. 2, pp. 135–188. Academic Press, New York.

7

Myogenesis: Differentiation of Skeletal Muscle Fibers

THOMAS L. LENTZ

I. Introduction

Skeletal muscle is formed by a process of differentiation from relatively unspecialized mononuclear cells. During this process, the cells develop the extremely complex cytological machinery that is related to the contractile function of muscle tissue. Proteins including actin and myosin are synthesized and polymerized into filamentous structures which become specifically oriented to one another in the myofibrils. At the same time elaborate membranous systems; the sarcoplasmic reticulum and the transverse tubular system, which are involved in the spread of excitation through the muscle fiber and the coupling of excitation and contraction develop in relation to the forming myofibrils. In skeletal muscle, the site of initial excitation, the motor end plate, arises in relation to the motor nerve fiber through structural and chemical specialization of the sarcolemma. The cytology of developing vertebrate skeletal muscle is presented in this chapter. Particular emphasis is placed on the differentiation of those cellular structures and organelles whose development results in acquisition of contractile activity.

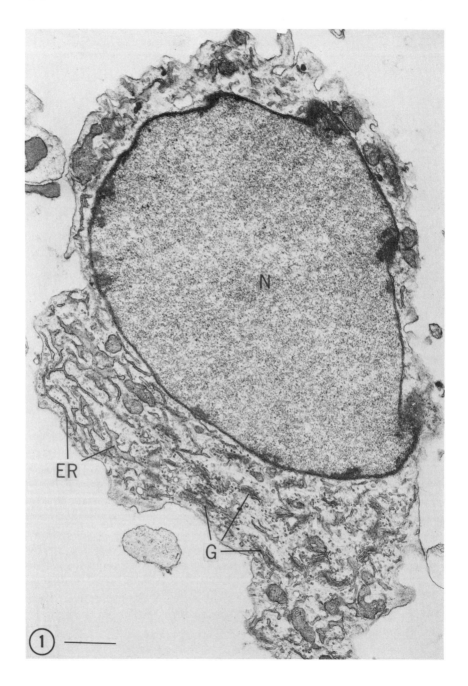

II. Structure of the Myoblast

In the embryo, the cells that give rise to muscle tissue are the myoblasts which originate from mesodermal mesenchymal cells (see Boyd, 1960, for review of embryonic development of muscle) (Fig. 1). Myoblasts are slightly elongated cells with a central nucleus (Fig. 2). The nuclear chromatin is dispersed. One or more nucleoli are present, composed of a central zone of fine-textured dense material surrounded by dense granules similar to ribosomes. Most striking is the abundance of ribosomes which are the prominent component of the cytoplasm (Hay, 1963; Heuson-Stiennon, 1965; Firket, 1967; Fischman, 1967; Lentz, 1969a; Shimada, 1971) (Fig. 2). They are arranged in clusters or clumps and in linear or curved chains. Membranous elements, on the other hand, are few in number. There are some short, scattered cisternae of rough-surfaced endoplasmic reticulum and a small Golgi apparatus near the nucleus. The abundance of polysomes and the paucity of rough-surfaced cisternae seem to distinguish myoblasts from mesenchymal cells (Fig. 1), fibroblasts, or other premyoblastic cells. The mitochondria are small and oval and have short cristae. Pinocytotic vesicles are associated with the plasma membrane of myoblasts.

The structure of myoblasts is typical of that of undifferentiated cells which generally have a large number of free ribosomes and few membranous specializations. Furthermore, the occurrence of free ribosomes is characteristic of cells that synthesize proteins for endogenous use (e.g., erythroblast). On the other hand, in cells that produce exogenous proteins, for example, the exocrine pancreas which releases its secretory product, the ribosomes are associated with an extensive endoplasmic reticulum. The proteins synthesized by the myoblast, of course, are retained within the adult cell although they become organized into complex fibrillar structures.

FIG. 1. Mesenchymal cell from the regenerating limb of the newt *Triturus*. Cells of this type are precursors of myoblasts and other cells including chondrocytes and fibroblasts. The cell is somewhat irregular in shape with a large nucleus (N). The cytoplasm is relatively specialized, containing rough-surfaced endoplasmic reticulum (ER) and Golgi complexes (G). The interconnecting cisternae of endoplasmic reticulum are dilated in places and contain moderately dense material. Golgi complexes are numerous and have a number of small vesicles associated with them. The hyaloplasm is of low density and contains mitochondria and some free ribosomes, especially in the interstices between cisternae. These cells are probably involved in the synthesis of intercellular matrix materials. (Bar represents 1 μm.) (From Lentz, 1969a.) ×14,000.

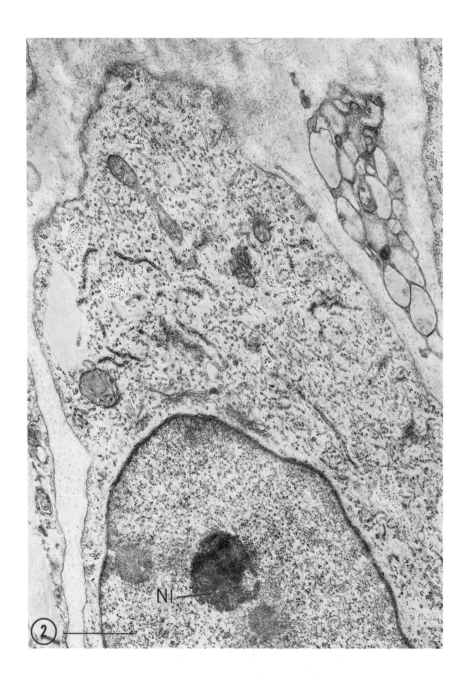

Myoblasts apparently can originate from sources other than the mesenchyme cells of the embryo. There is abundant evidence that in amphibians during regeneration specialized cells, such as muscle and cartilage, dedifferentiate giving rise to mesenchymal cells of the blastema (Butler, 1933; Thornton, 1938; Hay, 1962; Lentz, 1969a). Subsequently, the undifferentiated or mesenchymal cells differentiate to form muscle, bone, and connective tissues of the regenerating limb. The mesenchymal cells appear more complex than the myoblasts they give rise to. They have Golgi complexes and a system of intercommunicating cisternae of rough-surfaced endoplasmic reticulum (Fig. 1). These cells resemble fibroblasts and, in fact, may be secreting collagen precursors or other materials of the intercellular matrix (Hay, 1962).

Undifferentiated cells may persist beyond the early stages of development in the form of satellite cells. In adult muscle, satellite cells are small, flattened, and interposed between the plasma membrane and basement lamina of the muscle fiber. They are relatively uncomplicated in structure (Ishikawa, 1966). The nuclei are flattened and have condensed patches of chromatin. In the cytoplasm, a few channels of rough-surfaced endoplasmic reticulum are present, but free ribosomes are more abundant. Satellite cells have been found to incorporate precursors of DNA and to undergo mitosis. In growing animals, mitosis is followed by incorporation of one or both daughter nuclei into the adjacent muscle fiber (Moss and Leblond, 1970). They may also provide myoblasts for the regeneration of skeletal muscle in adult animals (Shafiq et al., 1967; Reznik, 1969).

III. Formation of the Multinucleated Myotube

There is general agreement that the multinucleated muscle fiber arises by fusion of mononucleate myoblasts (Lash et al., 1957; Capers, 1960; Konigsberg et al., 1960; Bergman, 1962; Betz et al., 1966; Shimada

---

FIG. 2. Myoblast from the regenerating limb of the newt. This mononucleate cell represents the earliest stage in muscle differentiation. The most conspicuous feature of the elongating cell are the free ribosomes which fill much of the cytoplasm and occur in clusters, clumps, or linear chains. A few rough- as well as smooth-surfaced cisternae of endoplasmic reticulum occur in the cytoplasm. Mitochondria are sparse and a Golgi complex occurs adjacent to the nucleus. The hyaloplasm is of low density and pinocytotic vesicles are numerous at the surface of the cell. The nucleus contains a conspicuous nucleolus (Nl) which is composed in part of granules similar to the cytoplasmic ribosomes. (Bar represents 1 μm.) (From Lentz, 1969a.) ×20,000.

et al., 1967). The cells with multiple, centrally located nuclei and with myofibrils at the periphery of the cytoplasm are known as myotubes (Boyd, 1960). Elongation of the cell, increase in number of myofibrils, and location of nuclei at the cell periphery produce the mature muscle fiber. The nuclei within the myotubes do not synthesize DNA, and so cannot divide mitotically or amitotically (Stockdale and Holtzer, 1961). Furthermore, when the embryonic muscle cells begin to synthesize the contractile proteins, they no longer replicate their DNA (Stockdale and Holtzer, 1961).

Proliferating myogenic stem cells may either divide forming new stem cells or following a critical division leave the mitotic cycle and become myoblasts which fuse to form myotubes (Bischoff and Holtzer, 1969). Initial union of myoblasts results in the formation of primary myotubes. Myoblasts clustered around the myotubes function as primordia of new generations of muscle cells which separate from the original myotube (Kelly and Zacks, 1969a). Thus, new generations of muscle differentiate by appositional growth along the walls of the large primary myotubes. Stem cells are gradually depleted during myogenesis, but some undifferentiated cells may persist on the surface of differentiated muscle fibers as satellite cells.

As emphasized by Bischoff and Holtzer (1969), fusion of myoblasts involves cell recognition and subsequent membrane–membrane interactions. Presumably, this is accomplished by the formation of specific macromolecular configurations on the surfaces of myoblasts and myotubes. Myoblasts fail to become incorporated into older myotubes indicating that a block to fusion develops. This block could reside in the basement membrane or lamina (Bischoff and Holtzer, 1969) which appears on the surface facing the mesenchymal compartment of myotubes and clusters of cells destined to become myotubes (Kelly, 1969).

In the actual process of fusion, the plasma membranes of adjacent cells break down in localized regions (Shimada, 1971) (Fig. 3a,b). A row of small vesicles and tubules is left along the boundary between the two cells (Hay, 1963; Shimada, 1971). Hay (1963) has suggested that the vesicles may be remnants of membranes that previously were discontinuous and that cell membranes disappear by breaking up into vesicles when the cells fuse. Many of the vesicles clearly are continuous with the original plasma membrane and some resemble pinocytotic vesicles (Fig. 3b).

The sarcoplasm of the resultant multinucleated cells shows areas with different degrees of differentiation (Lentz, 1969a; Shimada, 1971). This mosaic appearance is probably due to fusion of muscle-forming cells at different stages in their development. Fusion can occur between mono-

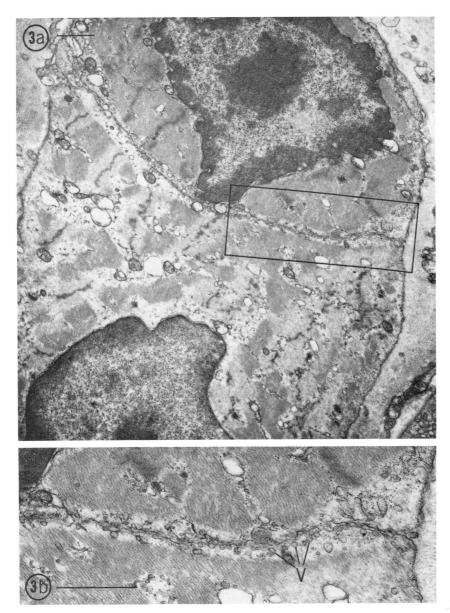

FIG. 3. (a) Two muscle cells in the process of fusion. The cells are well differentiated as evidenced by the abundance of myofibrils in the sarcoplasm. (Bar represents 1 μm.) ×10,000. (b) Higher magnification of the area enclosed by the rectangle in Fig. 3a. Along much of the zone of contact of adjacent cells the plasma membranes are apparent, but in some places they are indistinct. Small vesicles (V) are aligned in rows along the area of contact. (Bar represents 1 μm.) ×22,000.

nucleated myoblasts, between myoblasts and myotubes, and apparently between multinucleated myotubes (Shimada, 1971).

Specialized attachments have been observed between adjacent myotubes or between myoblasts and myotubes (Kelly and Zacks, 1969a; Shimada, 1971). Neighboring cells are separated over most of their surfaces by a space of 200 Å or more. In some places, attachment plaques resembling *fasciae adherentes* are found between adjacent cells (Shimada, 1971). These plaques consist of a symmetrical accumulation of electron-dense material in the cytoplasm adjacent to the plasma membranes. Elsewhere, myoblasts send a fingerlike projection of cytoplasm into adjacent cells.

In other regions, the membranes are closely apposed and approach one another to within 20 Å (Kelly and Zacks, 1969a). These regions, therefore, constitute close or gap junctions. In other regions, the membranes are fused obliterating the intercellular space to form tight junctions. Junctions of this type are rarely seen between mature skeletal muscle fibers and may have considerable importance in cell fusion and differentiation during development. Tight junctions function in cell to cell attachment (Farquhar and Palade, 1965) and could serve to bind together the differentiating muscle cells. These intercellular contacts may also mediate contact inhibition of cell movement (Trinkaus, 1969), thus connecting previously free or wandering cells to groups of cells with which they subsequently fuse.

Both gap and tight junctions also appear to be involved in intercellular transmission of ionic or other chemical signals (Loewenstein *et al.*, 1965; Furshpan and Potter, 1968; Bennett and Trinkaus, 1970; Lentz and Trinkaus, 1971). Thus, it has been suggested that the differentiating muscle cells are electrically coupled so that action potentials could pass between cells within groups permitting their coordinated activity (Kelly and Zacks, 1969a). Moreover, molecules larger than ions may also be capable of crossing these junctions (see Furshpan and Potter, 1968; Sheridan, 1968). In this case, information related to control or coordination of the developmental process could be exchanged between connected cells.

IV. Synthesis of Myofilaments

The bulk of myofilament synthesis seems to occur in multinucleated cells (Fischman, 1967; Lentz, 1969a; Shimada, 1971), although it does appear that synthesis of actin and myosin can begin in mononucleated myoblasts. Holtzer *et al.* (1957) and Engel and Horvath (1960) have shown that some mononucleated cells stain with fluorescein-labeled anti-myosin and presumably contain myosin. With the electron microscope, cells with apparently one nucleus are seen to have filaments (Fischman, 1967), although serial sections would be necessary to determine whether these cells are actually mononucleate.

At the time of appearance of myofilaments, the cytoplasm contains large numbers of ribosomal clusters (Figs. 4–6). These polysomes are large and composed of 50–80 ribosomes (Breuer et al., 1964; Heywood et al., 1967). The polysomes are generally considered to be the site of synthesis of actin and myosin (Hay, 1962; Breuer et al., 1964; Heuson-Stiennon, 1964; Allen and Pepe, 1965; Przybylski and Blumberg, 1966). In vitro, individual actin or myosin molecules polymerize into filaments (Huxley, 1964). Thus, it seems likely that individual actin and myosin molecules are synthesized on the polyribosomes and polymerize in the cytoplasm to produce filaments.

Free filaments, not closely associated with one another, are dispersed in the cytoplasm of developing myofibers. Two types of muscle filaments can be distinguished: thin filaments 50–70 Å in diameter and thick fila-ments 100–150 Å in diameter. Some investigators believe that thin fila-ments or actin appear prior to the thick myosin filaments (Ogawa, 1962; Price et al., 1964; Allen and Pepe, 1965; Obinata et al., 1966; Shafiq et al., 1967; Kelly, 1969) or that thick filaments appear before the thin filaments (Firket, 1967). Others feel that they appear simulta-neously (Hay, 1962, 1963; Dessouky and Hibbs, 1965; Przybylski and Blumberg, 1966; Fischman, 1967; Lentz, 1969a), although early in differ-entiation thin filaments greatly outnumber the thick (Fischman, 1967; Lentz, 1969a).

The thin filaments occur singly or in tangled skeins throughout the cytoplasm. They may contain bends or curves along their length. Some-times filaments overlie or are adjacent to clusters of ribosomes (Fig. 6). These unaligned filaments have the same diameter as the myofila-ments in the I band of developed myofibrils, and are presumed to be actin filaments.

The free thick filaments are straight or only slightly curved and have tapered ends. They possess lateral projections along their length. Thus, in size and structure they are the same as the filaments of the A band and correspond to myosin filaments. In the early myofiber, they do not seem to be as randomly distributed as thin filaments but are usually found near or as part of developing fibrils. Clusters of ribosomes are sometimes seen at the free ends of the thick filaments.

The close association of filaments with ribosomal clusters has been noted (Hay, 1963; Heuson-Stiennon, 1964; Przybylski and Blumberg,

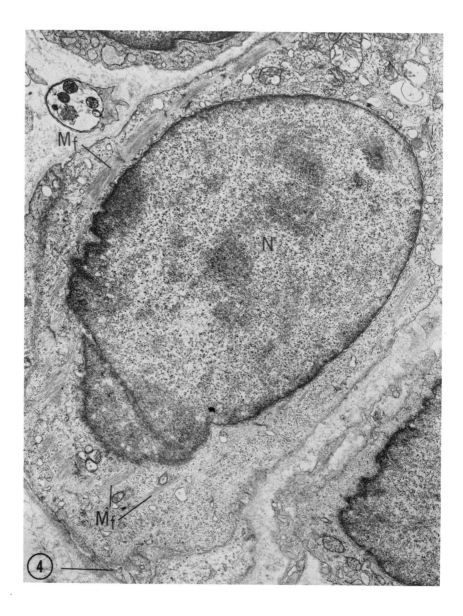

FIG. 4. Cell in early stage of myofibril formation. Like the myoblast, the cytoplasm has an abundance of free ribosomes. In addition, small myofibrils (Mf) have made their appearance in the sarcoplasm. The myofibril at the top of the cell is narrow but quite long with apparent striations. N, nucleus. (Bar represents 1 μm.) ×14,000.

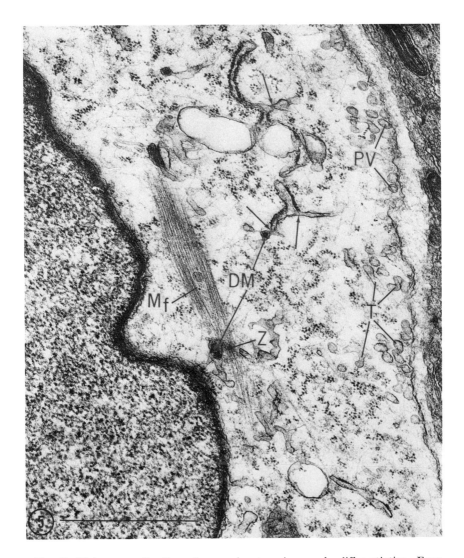

Fig. 5. Higher magnification of an early stage in muscle differentiation. From *Triturus*, 6-week regenerate. The cytoplasm contains clusters of free ribosomes and some free filaments. In addition, a small myofibril (Mf) is present. There are scattered, short cisternae of rough-surfaced endoplasmic reticulum which may be connected to smooth-surfaced tubules (arrows). Some of the smooth tubules contain masses of dense material (DM). One of these is situated immediately adjacent to the Z line of the myofibril. Pinocytotic vesicles (PV) are numerous in association with the sarcolemma. Just below the surface, short chains of vesicles form tubules (T) that represent the earliest stage of T-system formation. (Bar represents 1 μm.) (From Lentz, 1969a.) ×37,000.

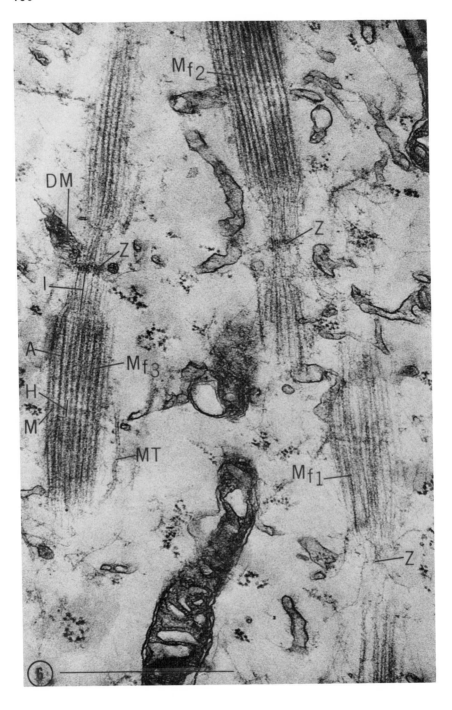

1966; Lentz, 1969a). Some investigators have interpreted this appearance as meaning that the filaments are spun out from the polyribosomes. It could also result simply from overlap of these structures in the section. Polymerization of actin and myosin molecules into filaments could take place at variable distances from the site of synthesis (ribosomes), depending on the extent of diffusion of the molecules.

In addition to the two populations of myofilaments, other filaments with a variety of diameters have been described. In some cases, these filaments or the thin filaments have been interpreted as representing stages in the polymerization of thick and thin filaments (Price et al., 1964; Heuson-Stiennon, 1965; Firket, 1967). Many of the variations in the reported sizes of filaments in the myoblast as well as discrepancies in the time of appearance of supposed myofilaments could be accounted for by the presence of a third population of filaments. These filaments have been observed in chick (Ishikawa et al., 1968) and newt (Kelly, 1969) myotubes and have a diameter of 100 Å. They are present in the myoblast and throughout myofiber development. They do not seem to be associated with formation of the myofibrils but instead may represent a cytoskeletal network (Kelly, 1969). In this regard, they bear close resemblance to the tonofilaments found in a variety of cell types.

Both the thick and thin filaments are oriented preferentially with the long axis of the muscle fiber, although the thin variety is more randomly oriented (Fischman, 1967). The longitudinal orientation of filaments is most pronounced in the subsarcolemmal regions of the cell. Microtubules are also abundant in these regions (Firket, 1967; Fischman, 1967). These structures are 200 to 240 Å in diameter and also oriented with the long axis of the cell. The microtubules are not connected with the myofilaments, but in view of their functions in other cells may determine the positioning of the filaments. Microtubules are thought to be involved in mediation of cytoplasmic streaming, movement of materials along their length, and shape determination of asymmetric cells (Ledbetter and Porter, 1963; Tilney and Porter, 1965; Bickel et al., 1966).

Fig. 6. Early stage in myofibril formation. Clusters of ribosomes and free filaments are abundant in the cytoplasm. A microtubule (MT) parallels one of the myofibrils. The myofibrils are composed of thick and thin filaments and in all of the fibrils, A, I, H, and M bands can be detected. In one myofibril (Mf₁), the region of the Z line is indicated only by the irregular convergence of thin filaments and does not contain dense material. In another fibril (Mf₂), there is less irregularity of thin filaments and a slight increase in density in the Z line region. In the third fibril (Mf₃), the Z line is denser and opposite it are membranous tubules containing dense material (DM) similar to that in the Z line. (Bar represents 1 μm.) (From Lentz, 1969a.) ×47,000.

If microtubules play a similar role in cytoplasmic streaming in muscle, they could be responsible for the longitudinal alignment of myofilaments by setting up cytoplasmic streams that orient the filaments (Fischman, 1967).

V. Differentiation of the Myofiber

A. MYOFIBRILS

Aggregation of free myofilaments into myofibrils must begin very shortly after synthesis of the filaments because cells containing free filaments usually have some small myofibrils (Fig. 4). Most of the newly formed myofibrils are found at the periphery of the myofiber beneath the sarcolemma. A major question concerns the presence or absence of striations in the earliest fibrils. Some investigators believe that the filaments aggregate directly into striated myofibrils (Bergman, 1962; Hay, 1963; Dessouky and Hibbs, 1965; Fischman, 1967; Shafiq et al., 1967; Lentz, 1969a). Others feel that unstriated fibrils are formed prior to the appearance of a banding pattern (Godman, 1957; Holtzer et al., 1957; Engel and Horvath, 1960; Holtzer, 1961; Wainrach and Sotelo, 1961; Allen and Pepe, 1965; Firket, 1967).

It seems likely that a certain number of filaments are necessary before it can be determined if they are oriented to one another in a specific manner. Thus, in the earliest myofibrils when the filaments are first becoming aligned and are loosely packed, it is not possible to discern the presence of transverse striations. However, once a certain number of filaments are present, it is apparent the two types are associated in a specific manner. Fischman (1967) has observed, for example, that in a myofibril containing only nine thick filaments, the filaments are packed in a hexagonal pattern. In longitudinal sections of small fibrils, it is apparent that the thin filaments overlap the thick filaments so that one end of the fibril is composed primarily of thin filaments and the other end of both thick and thin filaments (Fig. 6). Thus, although a definite banding pattern is not readily apparent in some of these small fibrils, the sites of the future A, I, and Z regions are already demarcated in early fibrils by the zone of thick and thin filaments, zone of thin filaments, and distal end of the thin filament zone, respectively.

The occurrence of striations in the early fibrils indicates that the banding pattern is a consequence of the nature of the interaction of the thick and thin filaments during their aggregation. The process of aggregation could be explained by specific interaction between the thick and thin filaments. If the monomeric units of actin and myosin filaments are oriented in the same direction to produce a structural polarity, interaction between the two protein molecules is likely to occur in only one critical orientation relative to each other (Huxley, 1963). In the myofibril, actin monomers in the thin filaments all point in one direction on one side of the Z line and in the opposite direction on the other side of the Z line. Furthermore, myosin molecules in the thick filaments are oriented in opposite directions in each half of the A band (Huxley, 1963). Since the molecules comprising the filaments are oriented, interaction between the filaments is also likely to be specific and polarized, and if one is reversed, the other also must be reversed (Huxley, 1963).

Actin and myosin molecules polymerize *in vitro* to form polarized filaments and the same process may occur *in vivo* (Huxley, 1963). Because of their polarization, the filaments will interact specifically during their aggregation into fibrils. Striations in the early fibrils may not be readily apparent at first because of incomplete interaction between filaments or the presence of only a small number of filaments. However, if the filaments interact specifically by virtue of their polarized nature, the banding pattern is a necessary consequence of their aggregation.

When the width of the fibril exceeds 0.1 to 0.2 μm, the typical banding pattern of mature fibrils can be discerned (Fig. 6). A and I bands are distinct, the H zone and M line are usually present, and the dense Z line is forming (see Section V,B). The length of the sarcomeres in newly formed fibrils is normal. Even in the early stages, it is common for the striations of separate myofibrils to be roughly in register. The myofibrils grow in length and width and become more numerous (Figs. 7, 8). At first, the numerous myofibrils seem to enlarge by the addition of filaments along their sides or at their ends. During the process of enlargement of the fibrils, the filaments become aligned in their characteristic positions with respect to adjacent filaments or the previously formed sarcomere.

B. Z Band

Perhaps the least well-understood aspect of myofibril formation is the development of the Z band, owing in part to the existence of different views on the architecture and chemical composition of the Z band (see Knappeis and Carlsen, 1962; Huxley, 1963; Franzini-Armstrong and Porter, 1964; Reedy, 1964; Kelly, 1967). Furthermore, there may be more than one mechanism for Z band formation, depending on the species. Heuson-Stiennon (1965) and Kelly (1969) have described the occurrence of Z bodies which coalesce to form Z bands. Z bodies are irregularly defined masses of dense material found at the periphery of myotubes, myoblasts, and their precursors. The Z bodies may be present prior to the appearance of actin and myosin filaments but are associated

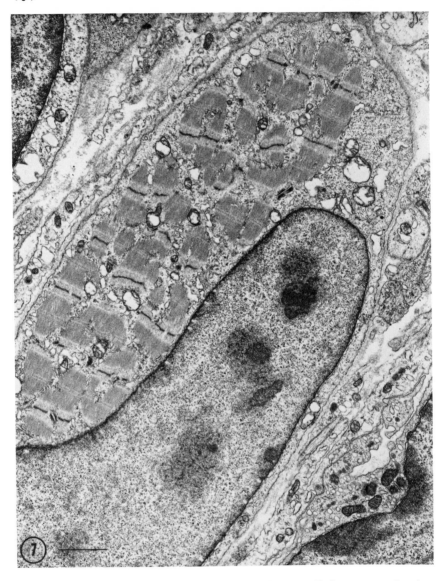

Fig. 7. Muscle cell in a later stage of differentiation. Myofibrils are more abundant and occupy much of the sarcoplasm. The striations of the fibrils are roughly in parallel. Elsewhere, ribosomes still occur in large numbers. The cell as well as its nucleus are elongated. (Bar represents 1 μm.) ×13,000.

with skeins of finely filamentous material. The latter has been interpreted as representing either precursors of actin filaments or as a nonactin, possibly tropomyosin, template upon which the actin filaments become

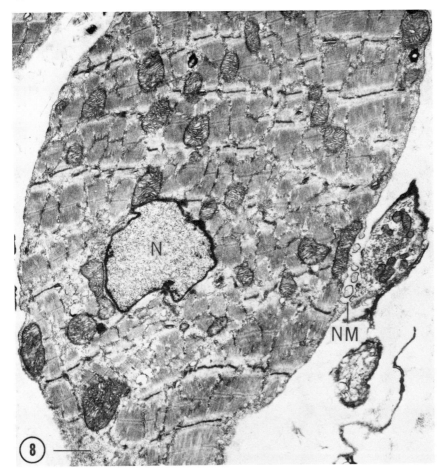

Fɪɢ. 8. Late stage in myofiber development. The myofibrils now fill most of the cytoplasmic space. The nucleus (N) is still central in position indicating the cell is not fully differentiated. A neuromuscular junction (NM) is seen on the surface of the fiber but does not cover an extensive area. The axon terminal is enveloped on its outer surface by Schwann cell cytoplasm and contains mitochondria and a large number of synaptic vesicles. Junctional folds on the muscle surface are well developed. (Bar represents 1 μm.) ×10,000.

attached (Kelly, 1969). Myosin filaments subsequently interact and become registered with the actin filaments attached to the Z bodies resulting in the appearance of A, I, and Z bands in the myofibril. Thus, according to this view, Z band material appears prior to other components of the myofibril and determines the subsequent interaction and alignment of the muscle filaments.

Other investigators have not observed early appearance of Z band material in the cytoplasm. Furthermore, although some early small myofibrils contain a dense Z band, others lack the dense material (Hay, 1963; Allen and Pepe, 1965; Firket, 1967; Lentz, 1969a). Even in the latter fibrils, the site of the Z band is indicated by the convergence or crossing of the ends of the thin filaments (Fig. 6). Thus, the interaction of the free ends of the thin filaments may form an irregular Z band lattice upon which the extremely dense material is subsequently deposited. In the newt, material identical in density and texture to that in the Z band is seen within the dilated terminal cisternae of the sarcoplasmic reticulum which become opposed to the developing Z line (Lentz, 1969a) (Figs. 5, 6). Firket (1967), on the other hand, has suggested the dense material lies within the transverse tubules. Since the sarcoplasmic reticulum is connected with the rough-surfaced endoplasmic reticulum, it is possible the dense material is synthesized by the rough-surfaced endoplasmic reticulum, segregated in channels of smooth endoplasmic reticulum, and then transported out of the smooth tubules at the Z band region. In this case, Z band material is deposited at the same time or following initial formation of myofibrils.

C. SARCOPLASMIC RETICULUM

In the earliest myotube stage, vesicular profiles and tubules of smooth-surfaced endoplasmic reticulum (sarcoplasmic reticulum) are scattered in the cytoplasm among ribosomes. Rough-surfaced endoplasmic reticulum is not abundant but occurs as short, isolated cisternae. Many images are seen where the smooth channels are continuous with the rough cisternae (Ezerman and Ishikawa, 1967; Lentz, 1969a; Schiaffino and Margreth, 1969) (Fig. 5). Thus, it has been suggested that the rough endoplasmic reticulum is the site of formation of the smooth sarcoplasmic reticulum membranes (Ezerman and Ishikawa, 1967). The earliest elaboration of sarcoplasmic reticulum seems to occur without an intimate association with myofibrils. The smooth tubules subsequently elongate, branch, and form connections with transverse tubules and myofibrils. The portion of the sarcoplasmic channels adjacent to T tubules is sometimes dilated and contains a mass of finely granular, dense material (Ezerman and Ishikawa, 1967; Lentz, 1969a; Schiaffino and Margreth, 1969).

D. TRANSVERSE TUBULAR SYSTEM

The origin and early development of the transverse tubular system, or T system, have been described by Ezerman and Ishikawa (1967) and Ishikawa (1968). Developing T-system tubules have been found in the earliest myotube stage. T tubules appear first as inpocketings and invaginations of the sarcolemma. The T system then develops as inward extensions of these invaginations. Large numbers of vesicles or caveolae resembling pinocytotic vesicles are associated with the surface of myoblasts and myotubes (Ishikawa, 1968; Lentz, 1969a) (Fig. 2). Near the surface and often connected with it are a number of elongated or branched channels (Fig. 5). These channels have a beaded or multilobed contour indicating they could form by repeated outpocketing from the original vesicles or caveolae (Ishikawa, 1968). Thus, the T-system tubule may be formed by a process of repeated caveolation of the surface tubules. The continuity of these structures with the cell surface is demonstrated by the observation that all the vesicular or tubular forms become filled with ferritin particles added to the extracellular medium.

Initially, the T-system tubules invaginate without any intimate relationship to sarcoplasmic reticulum or myofibrils. Sometimes the T tubules and sarcoplasmic reticulum intermingle in a complicated manner near the cell surface. T tubules can be observed deeper in the myotube as development proceeds and myofibrils become more abundant. Connections are observed between T-system tubules and sarcoplasmic reticulum and less commonly with rough-surfaced cisternae. The simplest connections are appositions of plasma membranes separated by a space of 100–200 Å. Later, more specialized connections between T tubules and sarcoplasmic reticulum are seen. These resemble those of mature muscle in which periodic densities occupy the intertubular space. Diads formed by apposition of a T tubule and one smooth-surfaced channel are frequent at first, but typical triads become more common. In amphibians, diads or triads may be found isolated in the sarcoplasm or in very close association with early myofibrils. In the psoas muscle of the rat, however, diads or triads are not positioned adjacent to the myofibrils until the fibrils are considerably well along in development (Schiaffino and Margreth, 1969). The fact that the latter fibrils are already in register would seem to argue against the possibility that the sarcoplasmic reticulum or T tubules play a role in the positioning of the myofibrils.

VI. Innervation of the Myofiber

All the cytological components of the mature muscle fiber, including myofibrils, sarcoplasmic reticulum, and T system, can develop in the absence of innervation (Shimada *et al.*, 1967). However, once muscle cells have acquired a nerve supply, they become dependent on it, and subsequently atrophy and degenerate following denervation (Tower, 1939). The motor nerve fibers do exert a morphogenetic effect on the formation of the motor end plate (Zelená, 1962; Teräväinen, 1968; Kelly and Zacks, 1969b; Lentz, 1969b); uninnervated cells do not develop cholinesterase(ChE)-containing motor end plates (Engel, 1961). Thus, the motor neurons appear to influence the chemical, structural, and functional specializations of the muscle fiber so that innervation should be considered an integral part of myogenesis.

During normal development, muscle differentiation progresses through the multinucleate stage containing myofibrils before neuromuscular junctions are formed (Hirano, 1967; Lentz, 1969b) (Fig. 8). Junctions develop only in relation to dilated intercellular nerve terminations that contain large numbers of small vesicles, 500 Å in diameter. These vesicles appear to be the same as those occurring in fully developed terminals and probably represent synaptic vesicles containing acetylcholine. Schwann cell cytoplasm, which at first completely envelops the nerve terminal, becomes discontinuous on the side of the axon termination facing the muscle surface. The first indication of end plate formation is the appearance of slight ridges or elevations on the muscle surface. There is a slight increase in density of the cytoplasm immediately beneath the plasma membrane of the elevations which can be identified as the first emergence of junctional folds. The axon becomes more closely approximated to the muscle surface and at the same time the surface ridges increase in length and constrict at their bases to form junctional folds (Figs. 9, 10). In the axon terminal, synaptic vesicles become concentrated in focal accumulations where the axon contour projects slightly opposite the secondary synaptic clefts between junctional folds. Mature neuromuscular junctions are formed by increase in the area of axon contact with muscle. Fully formed motor end plates are found on muscle fibers that are also fully differentiated.

ChE activity develops concurrently with morphological differentiation and first appears when axon terminations are closely approximated to the muscle surface and junctional folds are beginning to emerge (Fig. 9). Enzyme activity has not been found on intercellular nerve fiber endings or on the muscle surface prior to the appearance of morphological indications of motor end plate formation (Lentz, 1969b). This finding is in agreement with the conclusion that concentration of enzyme activity at the end plate region occurs only after axon contact with muscle has been established (Mumenthaler and Engel, 1961; Zelená, 1962; Khera and Laham, 1965; Lentz, 1969b). Other investigators, however, believe that ChE appears before innervation and suggest prior appearance of the enzyme may have a chemotactic influence on the approaching motor

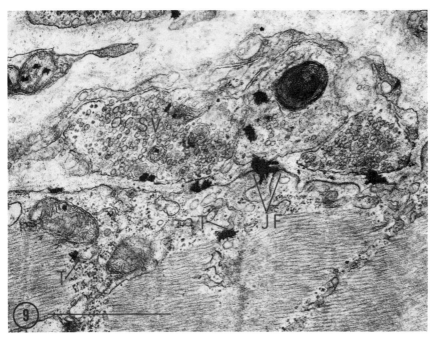

Fig. 9. Early neuromuscular junction in an 8-week amphibian regenerate stained for cholinesterase activity. The nerve is situated adjacent to the muscle cell and junctional folds (JF) are emerging. At this early stage, only a few lead deposits are present. Within the sarcoplasm beneath the junction, reaction product overlies two irregularly shaped membranous tubules (T). At the muscle surface, all the deposits seem to be associated with the outer surface of the sarcolemma, the largest situated at the tip of the most highly developed fold. In the nerve ending, lead is localized on the inner aspect of the membrane and also in the axoplasm. SV, Synaptic vesicles. (Bar represents 1 μm.) ×33,000.

nerve fiber (Kupfer and Koelle, 1951; Beckett and Bourne, 1958; Shen, 1958).

Acetylcholinesterase may be synthesized on ribosomes that persist in the end plate region (Padykula and Gauthier, 1970) while decreasing elsewhere in the muscle fiber. In the earliest stages of junction formation, ChE activity is first detected cytochemically in small tubulovesicles in the sarcoplasm in the region immediately beneath the axon (Lentz, 1969b) (Fig. 9). Some of these tubulovesicles are connected to the surface indicating they could have a role in delivery of the enzyme from deeper levels to the cell surface (Bloom and Barrnett, 1966; Lentz, 1969b). Larger deposits appear on the outer surface of the sarcolemma, indicating accumulation of enzyme at this site (Fig. 9). As morphological development of the neuromuscular junction progresses, the reactive sites on

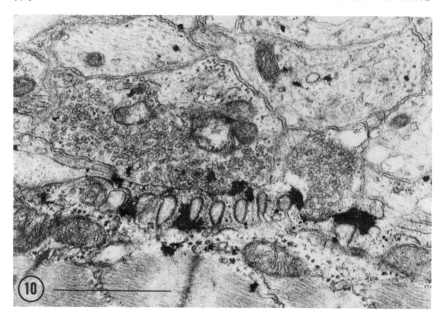

FIG. 10. Further stage in the development of the neuromuscular junction in the regenerating limb of the newt *Triturus*. Note that junctional folds are more numerous and elongated and comparable to those in the normal junction. Cholinesterase activity as demonstrated by the thiolacetic acid–lead nitrate method is more intense and localized primarily to the synaptic cleft. (Bar represents 1 μm.) ×32,000.

the plasma membrane increase in size and number (Fig. 10) until the final reaction product fills the synaptic cleft.

Both the appearance of junctional folds and ChE activity are modifications of the postsynaptic membrane that are involved in synaptic function. Their development only in relation to the motor axon supports the notion that the nerve is essential to the formation of both the morphological and chemical specializations of the neuromuscular junction.

References

Allen, E. R., and Pepe, F. A. (1965). *Amer. J. Anat.* **116,** 115.
Beckett, E. B., and Bourne, G. H. (1958). *Acta Anat.* **35,** 224.
Bennett, M. V. L., and Trinkaus, J. P. (1970). *J. Cell Biol.* **44,** 592.
Bergman, R. A. (1962). *Bull. Johns Hopkins Hosp.* **110,** 187.
Betz, E. H., Firket, H., and Reznik, M. (1966). *Int. Rev. Cytol.* **19,** 203.
Bickel, D., Tilney, L. G., and Porter, K. R. (1966). *Protoplasma* **61,** 322.
Bischoff, R., and Holtzer, H. (1969). *J. Cell Biol.* **41,** 188.

Bloom, F. E., and Barrnett, R. J. (1966). *J. Cell Biol.* **29**, 475.

Boyd, J. D. (1960). *In* "Structure and Function of Muscle" (G. H. Bourne, ed.), 1st ed., Vol. 1, pp. 63–85. Academic Press, New York.

Breuer, C. B., Davies, M. C., and Florini, J. R. (1964). *Biochemistry* **3**, 1713.

Butler, E. G. (1933). *J. Exp. Zool.* **65**, 271.

Capers, C. R. (1960). *J. Biophys. Biochem. Cytol.* **7**, 559.

Dessouky, D. A., and Hibbs, R. G. (1965). *Amer. J. Anat.* **116**, 523.

Engel, W. K. (1961). *J. Histochem. Cytochem.* **9**, 66.

Engel, W. K., and Horvath, B. (1960). *J. Exp. Zool.* **144**, 209.

Ezerman, E. B., and Ishikawa, H. (1967). *J. Cell Biol.* **35**, 405.

Farquhar, M. G., and Palade, G. E. (1965). *J. Cell Biol.* **26**, 263.

Firket, H. (1967). *Z. Zellforsch. Mikrosk. Anat.* **78**, 313.

Fischman, D. A. (1967). *J. Cell Biol.* **32**, 557.

Franzini-Armstrong, C., and Porter, K. R. (1964). *J. Cell Biol.* **22**, 675.

Furshpan, E. J., and Potter, D. D. (1968). *Curr. Top. Develop. Biol.* **3**, 95.

Godman, G. C. (1957). *J. Morphol.* **100**, 27.

Hay, E. D. (1962). *In* "Regeneration" (D. Rudnick, ed.), pp. 177–210. Ronald Press, New York.

Hay, E. D. (1963). *Z. Zellforsch. Mikrosk. Anat.* **59**, 6.

Heuson-Stiennon, J.-A. (1964). *J. Microsc. (Paris)* **3**, 229.

Heuson-Stiennon, J.-A. (1965). *J. Microsc. (Paris)* **4**, 657.

Heywood, S. M., Dowben, R. M., and Rich, A. (1967). *Proc. Nat. Acad. Sci. U.S.* **57**, 1002.

Hirano, H. (1967). *Z. Zellforsch. Mikrosk. Anat.* **79**, 198.

Holtzer, H. (1961). *In* "Synthesis of Molecular and Cellular Structure" (D. Rudnick, ed.), pp. 35–87. Ronald Press, New York.

Holtzer, H., Marshall, J. M., and Finck, H. (1957). *J. Biophys. Biochem. Cytol.* **3**, 705.

Huxley, H. E. (1963). *J. Mol. Biol.* **7**, 281.

Huxley, H. E. (1964). *Proc. Roy. Soc., Ser. B* **160**, 442.

Ishikawa, H. (1966). *Z. Anat. Entwicklungsgesch.* **125**, 43.

Ishikawa, H. (1968). *J. Cell Biol.* **38**, 51.

Ishikawa, H., Bischoff, R., and Holtzer, H. (1968). *J. Cell Biol.* **38**, 539.

Kelly, A. M., and Zacks, S. I. (1969a). *J. Cell Biol.* **42**, 135.

Kelly, A. M., and Zacks, S. I. (1969b). *J. Cell Biol.* **42**, 154.

Kelly, D. E. (1967). *J. Cell Biol.* **34**, 827.

Kelly, D. E. (1969). *Anat. Rec.* **163**, 403.

Khera, K. S., and Laham, Q. N. (1965). *J. Histochem. Cytochem.* **13**, 559.

Knappeis, G. G., and Carlsen, F. (1962). *J. Cell Biol.* **13**, 323.

Konigsberg, I. R., McElvain, N., Tootle, M., and Herrmann, H. (1960). *J. Biophys. Biochem. Cytol.* **8**, 333.

Kupfer, C., and Koelle, G. B. (1951). *J. Exp. Zool.* **116**, 397.

Lash, J. W., Holtzer, H., and Swift, H. (1957). *Anat. Rec.* **128**, 679.

Ledbetter, M. C., and Porter, K. R. (1963). *J. Cell Biol.* **19**, 239.

Lentz, T. L. (1969a). *Amer. J. Anat.* **124**, 447.

Lentz, T. L. (1969b). *J. Cell Biol.* **42**, 431.

Lentz, T. L., and Trinkaus, J. P. (1971). *J. Cell Biol.* **48**, 455.

Loewenstein, W. R., Socolar, S. J., Higashino, S., Kanno, Y., and Davidson, N. (1965). *Science* **149**, 295.

Moss, F. P., and Leblond, C. P. (1970). *J. Cell Biol.* **44**, 459.

Mumenthaler, M., and Engel, W. K. (1961). *Acta Anat.* **47,** 274.

Obinata, T., Yamamoto, M., and Maruyama, K. (1966). *Develop. Biol.* **14,** 192.

Ogawa, Y. (1962). *Exp. Cell Res.* **26,** 269.

Padykula, H. A., and Gauthier, G. F. (1970). *J. Cell Biol.* **46,** 27.

Price, H. M., Howes, E. L., and Blumberg, J. M. (1964). *Lab. Invest.* **13,** 1279.

Przybylski, F. E., and Blumberg, J. M. (1966). *Lab. Invest.* **15,** 836.

Reedy, M. K. (1964). *Proc. Roy. Soc., Ser. B* **160,** 458.

Reznik, M. (1969). *J. Cell Biol.* **40,** 568.

Schiaffino, S., and Margreth, A. (1969). *J. Cell Biol.* **41,** 855.

Shafiq, S. A., Gorycki, M. A., and Milhorat, A. T. (1967). *Neurology* **17,** 567.

Shen, S. C. (1958). *In* "The Chemical Basis of Development" (W. B. McElroy and B. Glass, eds.), pp. 416–432. Johns Hopkins Press, Baltimore, Maryland.

Sheridan, J. D. (1968). *J. Cell Biol.* **37,** 650.

Shimada, Y. (1971). *J. Cell Biol.* **48,** 129.

Shimada, Y., Fischman, D. A., and Moscona, A. A. (1967). *J. Cell Biol.* **35,** 445.

Stockdale, F. E., and Holtzer, H. (1961). *Exp. Cell Res.* **24,** 508.

Teräväinen, H. (1968). *Z. Zellforsch. Mikrosk. Anat.* **87,** 249.

Thornton, C. S. (1938). *J. Morphol.* **62,** 17.

Tilney, L. G., and Porter, K. R. (1965). *Protoplasma* **60,** 317.

Tower, S. S. (1939). *Physiol. Rev* **19,** 1.

Trinkaus, J. P. (1969). "Cells into Organs: The Forces that Shape the Embryo." Prentice-Hall, Englewood Cliffs, New Jersey.

Wainrach, S., and Sotelo, J. R. (1961). *Z. Zellforsch. Mikrosk. Anat.* **55,** 622.

Zelená, J. (1962). *In* "The Denervated Muscle" (E. Gutmann, ed.), pp. 103–126. Publ. House Czech. Acad. Sci., Prague.

8

Some Comparative Aspects of Cardiac and Skeletal Myogenesis

FRANCIS J. MANASEK*

I. Introduction

All mature muscle cells have a specialized intracellular, proteinaceous, contractile apparatus. Based on the appearance of these contractile elements, vertebrate muscle can be divided into two general groups, smooth and striated. In routine light or electron microscope preparations, smooth muscle, usually associated with involuntary movement, does not appear to have its contractile elements organized into orderly arrays of repeating subunits (see, however, Rosenbluth, 1971; Rice *et al.*, 1970). On the other hand, striated muscle demonstrates a highly ordered system of contractile structures, the myofibrils. The well-known periodicity of the myofibrillar subunits, the sarcomeres, is obvious at even the light microscope level and gives striated muscle its unique cross-banded appearance.

* The author's previously unpublished work included in this chapter was supported by U.S. Public Health Service Grant HE-10436 from the National Heart Institute. The author is a Medical Foundation Research Fellow.

193

Striated muscle falls into two distinct categories, skeletal and cardiac; both are capable of relatively rapid, forceful contractions and have their contractile proteins organized into discrete fibrils. The contractile myofibrils occupy the bulk of the mature muscle cell volume, yet both cardiac and skeletal muscle develop from undifferentiated precursor cells lacking these structures. The acquisition of fibrils is therefore the most striking and obvious event of striated muscle cytodifferentiation and certainly it has received the most investigative effort.

Striated muscle differentiation, however, is not limited solely to the acquisition of myofibrils. Elements of the sarcoplasmic reticulum must form and become properly aligned; cells may fuse, as in the case of skeletal myogenesis, to form a syncytium; unique cell-to-cell junctions form in cardiac muscle; and intermediary metabolism undergoes changes. Cardiac muscle becomes an oscillator capable of spontaneous membrane depolarization. Thus, myogenesis cannot be defined by any single parameter, but involves the development of the large repertoire of structures and functions by which mature muscle is characterized.

On the tissue level, myogenesis involves changes not only in the predominant cell type (muscle cells), but also in their relationship to other structures and cells such as true connective tissue cells (e.g., fibroblasts), nerves, and extracellular matrix such as ground substance and collagen. Muscle, in both mature and developing organisms, is a tissue with its constituents acting in concert. However appealing it may be to study development of isolated muscle cells, a full understanding of myogenesis must encompass this spectrum of interactions. Unfortunately relatively little is known, even at the descriptive level, about changes in cell populations, extracellular materials, and the interactions of these elements during myogenesis. Most of the available information concerns limb myogenesis and is discussed by Lentz in Chapter 7. Although his work will be alluded to in this chapter, I shall concentrate principally on events of cardiac myogenesis and attempt to explore comparative aspects of cardiac and skeletal myogenesis on the cell and tissue levels. Although there are significant differences between both mature and embryonic cardiac and skeletal muscle, many developmental studies appear to tacitly assume an essential similarity between these cell types during development. In this chapter, I shall try to demonstrate that these two cell types not only differ in their mature forms, but have significantly different developmental sequences on the cellular, tissue, and organ level.

Various authors use different names for myogenic cells in various developmental states. I will use the most common terminology for stages of skeletal muscle development; the term myoblast will denote a mononucleated cell that will fuse with other myoblasts (for a detailed discussion, see Lentz, Chapter 7) to form a multinucleated syncytium, the myotube. As the myotube matures, it acquires its full complement of fibrils and associated structures and is termed a myofiber, or skeletal muscle fiber. It should be noted that in the case of skeletal muscle the term myoblast simply means a mononucleated myogenic cell. It implies neither the presence or absence of myofibrils. The distinction between a presumptive myoblast and a myoblast will not be made. In the case of cardiac muscle, which does not become syncytial, the term myoblast will be used to refer to myogenic cells that do not yet contain organized myofibrils. Embryonic myocardial cells with fibrils are simply called "developing myocytes." Although there are other parameters by which muscle differentiation can be defined, such as chemical demonstration of specialized proteins (actin and myosin) or onset of contractility or irritability, I will deal largely with cytodifferentiation as determined by appearance of myofibrils. Many studies have attempted to elucidate the sequence of development of organized myofibrils (see Fischman, 1970; Lentz, Chapter 7) and it is generally assumed that similar physicochemical mechanisms underlie myofibrillogenesis in both cardiac and skeletal muscle.

In the following sections I shall describe the embryonic origin of skeletal and cardiac muscle and their course of embryonic development and stress the differences between them at various stages of maturation.

II. Embryonic Origin of Skeletal Muscle

The immediate precursors of multinucleated syncytial skeletal muscle myotubes are almost certainly mononucleated cells (for review, see Lentz, Chapter 7 and Fischman, 1970). The process by which these cells fuse to form syncytia has recently been examined at the electron microscope level using chick skeletal muscle cells grown *in vitro* (Shimada, 1971). A general review of the origin and development of myogenic precursors is presented by Boyd (1960).

A. Origin of Limb Skeletal Muscle

In the chick, skeletal muscle myoblasts of the appendage (limb) and portions of the trunk muscle are secondary mesenchyme (Hay, 1968) derived from dispersion of the secondary epithelium of the lateral plate. These mesenchymal cells migrate freely and cells from the same lateral plate origin probably also give rise to connective tissue cells. It is highly likely that during the initial migration of limb bud precursors, fibroblasts, chondroblasts, and myoblasts are in intimate relationship with each other and are intermingled although migration does not play a role in later development (Searls, 1967). The appearance of discrete regions in the developing limb and the problems involved in identifying myogenic from chondrogenic regions is discussed by Searls in Chapter 9. Since the bulk of the cells found in early limb development are mesenchymal, it is not possible in early stages to unequivocally distinguish these different cell types by morphological criteria alone, except, as Lentz discussed, myoblasts probably have a sparser granular endoplasmic reticulum. In culture, cells with myogenic potential can be operationally distinguished from "fibroblastic" cell types (Konigsberg, 1963).

B. Origin of Axial (Trunk) Musculature

It is probable that in the chick myogenic regions of the somites largely give rise to dorsal components of the axial skeletal system (Straus and Rawles, 1953; however, see Parry, 1968). The somites which are metameric, transient embryonic structures are true epithelia (Williams, 1911) with their constituent cells demonstrating apical junctional complexes along the adluminal surface (Hay, 1968). The myotome is also epithelial and in addition appears to be relatively homogeneous. This seems especially true in the larval teleost (Waterman, 1969) where it appears particularly obvious that somitic muscle does not come into intimate contact with connective tissue cells during early development.

III. Embryonic Origin of Cardiac Muscle

Cardiac myoblasts comprise a portion of the splanchnic mesoderm that is derived from the splitting of the lateral mesoderm into two components; the somatic and splanchnic layers. The splanchnic mesoderm, which is a secondary epithelium (Hay, 1968), forms two bilateral folds that come together at the embryonic midline. They retain their epithelial organization as they fuse to form the myocardial layer. Seemingly, at no time after the establishment of the splanchnic mesoderm do cardiac myoblasts become mesenchymal. This is an important point since the maintenance of the integrity of the myocardial epithelium may be essential for normal development; therefore, models of cardiac myogenesis based entirely on studies of dissociated cultured cardiac muscle should be used with caution.

Similar to the somite myotome, developing cardiac muscle exhibits a definitive apical surface with apical junctional complexes and a distinct basal surface with a basal lamina (Manasek, 1968b, 1970a) (Fig. 1). The epithelial nature of the splanchnic mesoderm is retained during its development into the myocardium which can be considered an epithelial tissue throughout development (Manasek, 1968b, 1970a; Hay, 1968). Rosenquist and De Haan (1966) were able to show by means of radioautographic tracing of grafts labeled with [³H]thymidine that cell mixing does not occur in the developing myocardium and that cells maintain their positions relative to one another during cytodifferentiation and organogenesis.

The developing myocardium is a pure population of developing myocytes (Manasek, 1968a,b, 1969b, 1970a) and initially contains no connective tissue cells. This point will be discussed in greater detail, but it should be stressed that the early heart contains only two cell types: developing myocytes of the myocardium and endocardium lining the lumen.

IV. Fibril Formation and Mitosis during Myogenesis

In addition to Lentz's chapter in this volume, two other reviews of molecular aspects of muscle differentiation and myofibril assembly have recently appeared (Herrmann et al., 1970; Fischman, 1970). Therefore, in this section I shall again stress differences in development of cardiac and skeletal muscle and, in particular, discuss aspects of cardiac muscle development that do not seem to have counterparts in skeletal myogenesis.

A. FIBRILLOGENESIS AND MITOSIS IN SKELETAL MUSCLE

In developing limb musculature, mononucleated skeletal myoblasts appear to proliferate by mitotic division, then after a quantal mitotic event stop dividing (Ishikawa et al., 1968). Myoblasts fuse to form syncytial myotubes which then form large numbers of myofibrils. In the intact developing limb, events of skeletal myogenesis occur over a relatively long period of time, and at most developmental stages both myotubes with fibrils and undifferentiated presumptive myoblasts can be demonstrated. Thus, limb skeletal muscle formation is asynchronous and a proliferative population of stem cells (myoblasts) remains, pre-

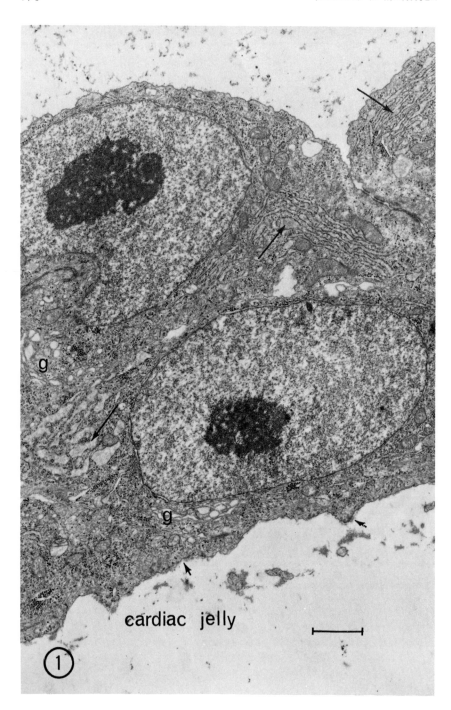

cardiac jelly

sumably dividing and providing additional cells for the growing limb muscle.

During early development in the chick, somite myoblasts contain material that binds fluorescent antimyosin antibody as early as stage 15 (Holtzer *et al.*, 1957). Overt fibrillogenesis occurs at about stages 16–17 with larger numbers of both thick and thin filaments seen in close association by about stages 18–20 (Holtzer *et al.*, 1957; Allen and Pepe, 1965; Przybylski and Blumberg, 1966). In both the chick (Przybylski and Blumberg, 1966) and the zebra fish (Waterman, 1969) fibrillogenesis occurs in mononucleated myoblasts. As can be seen in Figs. 2 and 3 these are not rudimentary fibrils, but contain well-demarcated A, I, and Z bands. Although such well-formed fibrils have not been reported in mononucleated limb skeletal muscle progenitors they should not be considered an "exception." Rather, it should be recognized that somitic and appendicular muscle have different developmental sequences.

Thus it appears that fibrillogenesis in epithelium-derived skeletal muscle is not necessarily linked to myotube formation as it largely seems to be in mesenchymatous limb muscle. Since many studies on skeletal muscle differentiation *in vitro* have been done using limb muscle it would seem particularly fruitful to examine somite muscle under similar conditions. For example, it remains to be established if the close association of myoblasts characteristic of the myotome *in vivo* (Waterman, 1969) is necessary for cytodifferentiation or if dispersed somitic myoblasts can also form fibrils.

Once incorporated into multinucleated myotubes, nuclei no longer divide (Stockdale and Holtzer, 1961), and the temporal correlation between cessation of division and appearance of fibrils during skeletal myogenesis has led some workers to generalize about differentiation and mitotic inhibition. The dichotomy between differentiation and mitosis has been noted also in electron microscope studies of developing somite muscle (Przybylski and Blumberg, 1966; Waterman, 1969) in which well-developed myofibrils occur in mononucleated cells. Although a decrease in DNA polymerase can be correlated with myoblast fusion (O'Neill and

---

FIG. 1. The epithelial characteristics of the developing myocardium are clearly demonstrated in this micrograph through the ventricular wall of a stage 10 (10-somite) chick embryo. A prominent apical junctional complex is seen in the upper right near the free surface (top) and a patchy basal lamina (small arrows) is forming between the basal (lower) surface and the cardiac jelly. Although developing myofibrils are present they are not prominent at this low magnification. However, the abundance of granular endoplasmic reticulum (arrows) and Golgi complexes (g) are prominent. The myocardium has the appearance of a secretory epithelium. The mark represents 1 μm.

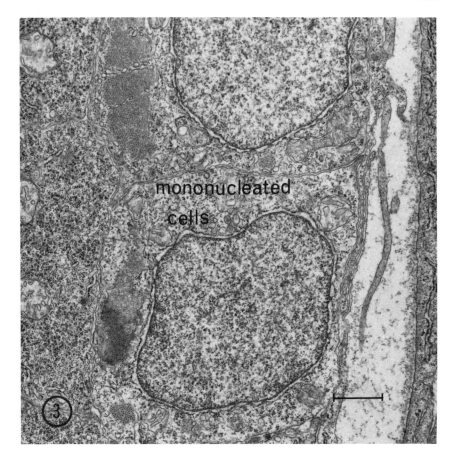

FIG. 3. This electron micrograph shows a cross section of a portion of a somite from a stage similar to that illustrated in Fig. 2. The conspicuous myofibrils within mononucleated skeletal muscle cells are not "exceptions," but seem character-istic of somite-derived axial muscle and are not noted in limb musculature derived from mesenchyme leaving the lateral plate. The mark represents 1 μm. Micrograph courtesy of Dr. Robert Waterman.

Strohman, 1969) and may account for cessation of DNA synthesis in myotubes, the events contributing toward mitotic inhibition in somite myoblasts remain largely unexplored.

FIG. 2. Mononucleated somite myoblasts are able to form well-developed myofibrils. In this example from the zebrafish larva, the most lateral cells of the somite (arrows, inset) are mononucleated yet they have distinct myofibrils as seen in the electron micrograph. The deeper cells (left side) are multinucleated. The mark on the electron micrograph represents 1 μm; that on the inset represents 10 μm. Micrographs courtesy Dr. Robert Waterman.

B. FIBRILLOGENESIS AND MITOSIS IN CARDIAC MUSCLE

The heart is very precocious in its development and in most organisms the onset of cardiac contractility occurs very early in embryonic life. For example, in the chick cardiac myoblasts become functional at about stage 10 and filaments are seen somewhat earlier.

The pattern of growth and differentiation of the myocardium is in sharp contrast to the events of limb skeletal muscle development. Cardiac myoblasts, as we have seen, are epithelial rather than mesenchymal; they do not fuse to form syncytia but maintain their cellular integrity, differentiate, and form myofibrils while they are mononuclear cells. Thus in this respect, cardiac myogenesis appears somewhat similar to early somitic skeletal myogenesis.

Shortly before the onset of visible myocardial contractions (in the chick at about stage 10) evidence of fibrillogenesis can be seen (Olivo *et al.*, 1964; Manasek, 1968b; Orts-Llorca and Gonzáles-Santander, 1969) and by the time spontaneous contractility occurs, cross-banded fibrils extending the entire length of the cells can be demonstrated (Manasek, 1968b). Moreover the onset of fibrillogenesis occurs rapidly throughout the myocardium and very shortly all the ventricular cells appear to have fibrils in various states of assembly. Cardiac muscle thus exhibits a relatively high degree of synchrony of myoblast–myocyte transformation. During this early differentiative period no cell types other than muscle are present (Manasek, 1968b, 1970a). Nonmuscle cells such as vascular constituents, fibroblasts, and epicardial cells normally seen in mature myocardium are added to the heart later in development (Manasek, 1969b, 1970a).

Early embryonic myofibrils are exceedingly difficult, and in many cases impossible, to detect with routine light microscope histological techniques. However, more and more fibrils become visible in increasingly large numbers of cells as the embryo matures (Fig. 4). Thus, it has been generally assumed that the early myocardium contains a mixed cell population, consisting of developing myocytes and undifferentiated myoblasts. These latter cells were thought to constitute a proliferative, or stem cell line, and that when they differentiated they were withdrawn from the mitotic pool. However, more recent electron microscope studies of cardiac myogenesis in the chick (Manasek, 1968b, 1969b, 1970a; Weinstein and Hay, 1970) could find no evidence for the persistence of undifferentiated myoblasts. Instead, the entire myocardial wall appears as a homogeneous tissue containing only developing myocytes.

Obviously the heart increases in both size and number of muscle cells between myoblast–myocyte transformation and the end of the embryonic

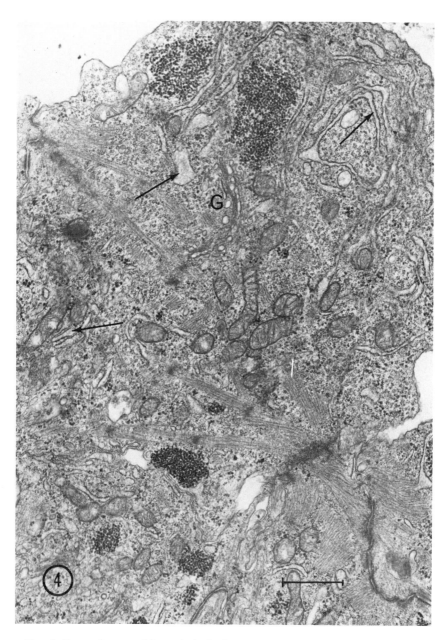

Fɪɢ. 4. Even after acquiring relatively large numbers of myofibrils, cardiac myocytes retain an extensive granular endoplasmic reticulum (arrows). This section from the heart of a 19-somite chick embryo shows the apparent lack of order in myofibrils forming in cardiac muscle. A Golgi complex (G) is prominent, as are pools of glycogen and developing intercalated discs. An apical junctional complex seemingly undergoing transformation to an intercalated disc is seen in the upper left. The mark represents 1 μm.

period, and the mitotic index (Grohmann, 1961) and the thymidine labeling index (Sissman, 1966; Jeter and Cameron, 1971) have been determined at different developmental stages. If there is no undifferentiated stem cell line in the myocardium, are these mitoses occurring in differentiated muscle? For the reasons discussed above it is not possible to demonstrate myofibrils in the dividing cells with the light microscope. However the electron microscope unequivocally demonstrates a well-developed complement of fibrils in these dividing myocytes (Manasek, 1968a; Rumyantsev and Snigirevskaya, 1968; Weinstein and Hay, 1970; Fischman, 1970). Thus despite virtually unimpeachable evidence that mitosis in skeletal muscle ceases prior to sythesis of fibrils (for discussion, see Herrmann *et al.*, 1970), this clearly is not the case in cardiac muscle (Fig. 5).

Radioautographic evidence suggests that differentiated cardiac muscle continues to replicate DNA. Weinstein and Hay (1970) clearly demonstrated nuclear labeling with [³H]thymidine in cardiac myocytes that contained well-formed fibrils. Jeter and Cameron's (1971) light microscope analysis indicated that 98% of all the cells in the intact 7-day heart incorporate [³H]thymidine (Table I). Whether all these cells are capable of entering mitosis remains problematical, but certainly some do since cells containing fibrils have been seen in mitosis as late as 11 days of incubation (Manasek, 1968a; see also Fig. 5). It is of course possible that some of the [³H]thymidine label represents endoreplication leading to increased DNA content of individual nuclei. Photometric studies on human cardiac muscle nuclei (Kompmann *et al.*, 1966) indicate that in mature normal hearts DNA content can range up to 16-ploid, and the percentage of high ploidy increases in hypertrophied hearts. However, it does not seem likely that such increases in DNA reflect an increase in total gene dosage (Bloom and Egli, 1969) because the maximum number of potential nucleolar organizers remains fixed. This problem as it relates to embryonic cardiac myogenesis is important and would be useful as a starting point to determine ploidy at various stages of development. Nonetheless, since electron microscope observations indicate that differentiated cardiac myocytes can and do divide, and radioautographic evidence indicate that they can synthesize DNA, we must

---

Fig. 5. Chromosomes and myofibrils are prominent in this differentiated cardiac myocyte fixed while undergoing mitosis. The presence of specialized structures, in this case myofibrils, is not necessarily incompatible with cell division. The arrows point to possible secretory granules that are retained during mitosis. This is a typical example of dividing myocytes found in the ventricle of an 8-day chick embryo. The mark represents 1 μm.

TABLE I

INCORPORATION OF [³H]THYMIDINE IN INTACT HEART

Total length of exposure to isotope	Labeling index (%) for Hamburger–Hamilton stage[a]							
	12	18	23	24	32	35	41	46
45 min	10	31	42	—	30	15	—	7
	12	—	30	—	—	19	—	—
	—	—	—	—	—	15	—	—
12 hr	45	89	97	—	—	49	—	4
	—	—	—	—	—	39	—	—
24 hr	—	—	—	—	—	52	27	15
48 hr	—	—	96	97	98	—	68	—
	—	—	—	96	—	—	—	—
	—	—	—	97	—	—	—	—

[a] In this table mean labeling index (determined by a minimum of three counts per radioautograph) of developing cardiac muscle cells at different Hamburger–Hamilton stages is plotted as a function of total time that embryos were incubated (*in ovo*) with [³H]thymidine. The 45-min group were sacrificed 45 min after a single administration of [³H]thymidine. The 12-hr group received three equally spaced administrations; the 24-hr group four administrations; the 48-hr group received a total of eight administrations of tritiated thymidine. Each number represents the average labeling index for a separate embryo. (Table compiled from selected data in Jeter and Cameron, 1971.)

reject the generalized concept that fibril synthesis and mitosis are mutually exclusive.

More recently Chacko (1971) presented radioautographic evidence for the existence of a postmitotic population of cardiac myocytes obtained from 5-day-old chick hearts trypsinized and grown *in vitro*. This finding correlates well with results obtained *in vitro* by Rumery and Rieke (1967) who also were unable to demonstrate [³H]thymidine uptake by cells with cross-striated myofibrils. These results appear in contradiction to those of Jeter and Cameron (1971) (see Table I), who measured [³H]thymidine incorporation *in vivo*. Although this apparent contradiction is yet to be explained, the mitotic block seen *in vitro* may be a result of the culture environment. The problem remains unsolved and we again must ask the question: How is cell division regulated in the intact developing heart?

It still remains possible that extrinsic factors may regulate mitosis in the intact developing organ (Manasek, 1968a). Thus, changes in the extracellular matrix which constitutes the immediate environment may conceivably play a regulative role in organ growth. Although we will have a little to say about the origin of embryonic cardiac connective tissue (extracellular matrix) in the next section, virtually nothing is known about its composition at different times in development or about possible interactions of the matrix with developing cells and tissues of the heart. This appears to be a potentially fertile field and one that in principle lends itself well to rigidly controlled *in vitro* experiments.

V. Other Cell Functions during Myogenesis

A. SECRETORY ACTIVITY OF CARDIAC MUSCLE

We noted earlier that cardiac muscle is an epithelium. In the intact developing heart it rests on a layer of embryonic connective tissue called cardiac jelly (Davis, 1924), separating myocardium from endocardium (Fig. 6). Histochemical techniques (Ortiz, 1958) and biochemical analyses (Gessner *et al.*, 1965) indicate that cardiac jelly is an acid muco-polysaccharide probably containing chondroitin sulfate. An early radio-autographic investigation (Johnston and Comar, 1957) suggested that the cardiac jelly was produced by the endocardium. However, a more recent radioautographic study using embryos incubated *in vitro* with $^{35}SO_4{}^{2-}$ for relatively brief periods strongly implicated the myocardium as a source of cardiac jelly (Manasek, 1970b). Radioautographs of sections through the heart of a 12-somite embryo incubated with $^{35}SO_4{}^{2-}$ for 1 hr 40 min (Fig. 6) demonstrate intense label over the myocardium and the cardiac jelly. With shorter incubation periods, label was primarily situated over the myocardium and progressively longer incubations resulted in increasingly heavy extracellular label (Fig. 7). This sequence has now been established for the developmental range encompassing stage 8 to 13 (Manasek, 1970b, and unpublished work). Thus, with respect to sulfated extracellular material (likely chondroitin sulfate) both cardiac myoblasts (in the preheart embryo) and functional myocytes are secretory. A continuation of these studies has demonstrated that sugars such as [^{3}H]glucosamine (Fig. 8) and galactose are also utilized by the myocardium for extracellular matrix production during this developmental range. At no time was any radioautographic evidence obtained, by using either $^{35}SO_4{}^{2-}$ or tritiated sugars, implicating the endocardium as a source of cardiac jelly in young embryos. The implication of cardiac myocytes in the elaboration of cardiac connective tissue is an important point since secretion of matrix is a function usually ascribed to "fibroblasts" that are assumed to be specialized for this function (Thorp and Dorfman, 1967).

Recently, A. Cohen (unpublished work) has demonstrated that the amino acid proline is also taken up by developing myocardium and

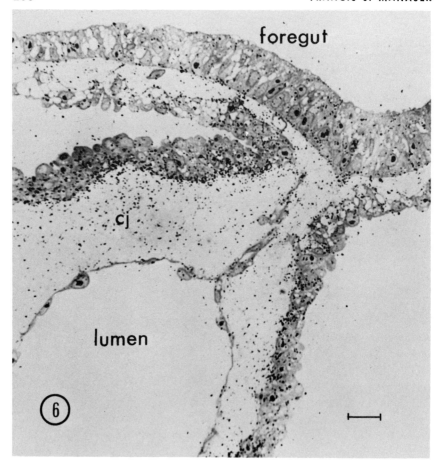

Fig. 6. This light microscope radioautograph shows a portion of the heart of a 12-somite embryo that had been incubated with $^{35}SO_4^{2-}$ *in vitro* for 1 hr 40 min. Label is extremely heavy over the myocardium and cardiac jelly (cj) nearest the myocardium. With short incubations the label is never concentrated in the sub-endocardial zone. The mark represents 25 μm.

utilized in the production of an extracellular material (Fig. 9). This preliminary result is highly suggestive of the possible production of collagen and tentatively similar to earlier indications that the myocardium may produce collagen (Gaines, 1960; Woessner *et al.*, 1967). Apparently cytodifferentiation does not inhibit any of these secretory functions. Moreover, an examination of the ultrastructure of the developing myocardium reveals a well-developed granular endoplasmic reticulum and extensive Golgi regions (Manasek, 1968b). Cultured cardiac muscle also contains a large amount of granular endoplasmic reticulum that apparently increases in amount if the incubation temperature is raised (Weissenfels, 1968). *In vivo*, the secretory organelles of developing myocardium are sufficiently extensive that the tissue has the distinct appearance of a secretory epithelium (Figs. 1, 4).

So far we have discussed secretion by muscle only in terms of production and elaboration of extracellular matrix. There is morphological evidence that cardiac muscle has another secretory role. Mature cardiac myocytes (especially atrial cells) contain membrane-limited electron-dense granules that appear to be produced in the Golgi apparatus (Jamison and Palade, 1964). No comparable structures have been reported in skeletal muscle. Largely because of their morphological appearance, it has been suggested that these granules contain catecholamines. Although it remains to be proved, their apparent synthesis within Golgi systems is strongly suggestive that these granules represent a secretory product. Similar granules have been seen in embryonic rat heart (Pager, 1968) and embryonic chick ventricle (Manasek, 1969a; Weinstein and Hay 1970; see Fig. 10). Thus, if the presence of these granules represents a true secretory activity (possibly catecholamine elaboration), then this function begins at a relatively early stage when the myocytes are still synthesizing myofibrils.

B. Secretory Activity of Skeletal Muscle

Developing skeletal muscle is usually described as containing very little granular endoplasmic reticulum (see Fischman, 1970; Lentz, Chapter 7, this volume for discussions). These observations strongly suggest that skeletal myoblasts and myotubes do not elaborate large quantities of extracellular material. However, we must bear in mind the observations discussed by Searls that myogenic regions of early limb buds do incorporate inorganic ^{35}S which suggests that at some stage in development limb skeletal muscle may also pass through a secretory phase. At any rate, it seems as though this possible secretory phase is transient in limb skeletal muscle and can no longer be demonstrated by the time cytodifferentiation occurs (Medoff, 1967). Because of the heterogeneity of developing limb skeletal muscle it would appear that cloned cultures might provide the best material for an investigation of possible secretory activity of these cells.

Although somites *in vivo* have been shown to contain enzymes necessary for mucopolysaccharide synthesis (Lash, 1968) it is not clearly known in which part of the somite these enzymes are localized. Chondroitin sulfatelike material has been extracted from chick embryo

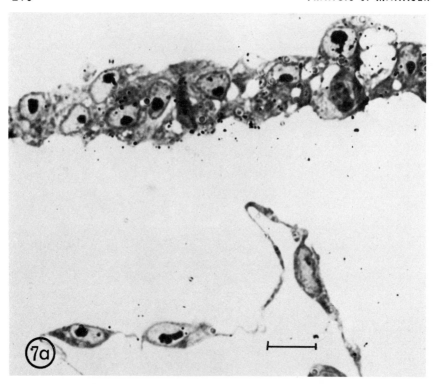

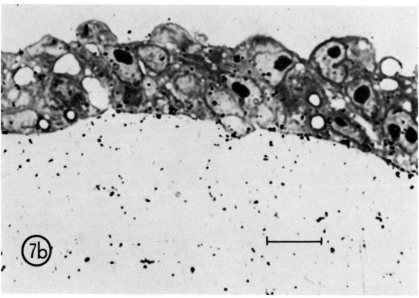

somites (Franco-Browder *et al.*, 1963), but again, the localization of this material in sclerotome, myotome, or dermatome was not made. Thus it cannot be stated at this time whether or not somitic skeletal muscle produces an extracellular matrix. In his study of somite myogenesis in the zebra fish, Waterman (1969) reported that there is very little intercellular space in the myotome and that formation of the septa and recognizable collagen fibrils occurred at about the time of hatching and concomitant with the appearance of mesenchymal connective tissue cells. It appears that there is only a small amount of extracellular material within the myotome at earlier stages.

VI. Conclusion and Summary

Cardiac and skeletal muscle have clearly discernable differences throughout their developmental history, and some of those discussed in this chapter are summarized in Table II.

Developing cardiac muscle may be particularly well suited for studies of the relationship of cell division to differentiation. Cardiac muscle is more typical of cells in general in that it is not a syncytium. In the skeletal muscle systems most often studied, the formation of syncytial myotubes, as well as fibril formation, can be correlated with cessation of cytokinesis and Weinstein and Hay (1970) point out that this former event, rather than fibrillogenesis may repress cell division. Developing cardiac muscle undergoes no events comparable to skeletal myoblast fusion. Moreover, O'Neill and Strohman (1969) showed that myoblast fusion *in vitro* could be closely correlated with a decrease in DNA polymerase. Unfortunately similar measurements of levels of enzymes necessary for DNA synthesis have not been made in cardiac muscle. It would be extremely interesting to determine if, in developing cardiac muscle, thymidylate kinase or DNA polymerase levels decrease in a fashion similar to the noted decrease in overall mitotic index (Grohmann, 1961) as the embryo matures. Similar studies would be valuable in the case of somite skeletal muscle since these mononucleated myoblasts de-

---

FIG. 7. A 1-hr incubation with $^{35}SO_4^{2-}$ results in label concentrated over the myocardium, with only light cardiac jelly label (a). If the embryo is incubated with $^{35}SO_4^{2-}$ for $1\frac{1}{2}$ hr the myocardium is labeled, but, in addition, significant numbers of silver grains are seen over the cardiac jelly (b). This sequence is further evidence that the myocardium is producing a sulfated extracellular matrix. The marks represent 20 μm.

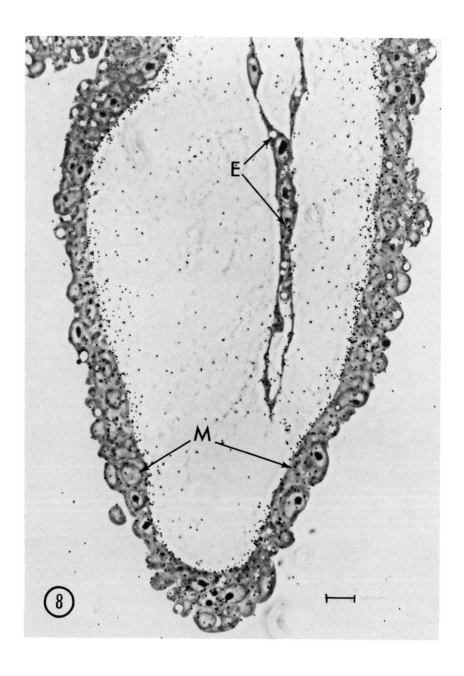

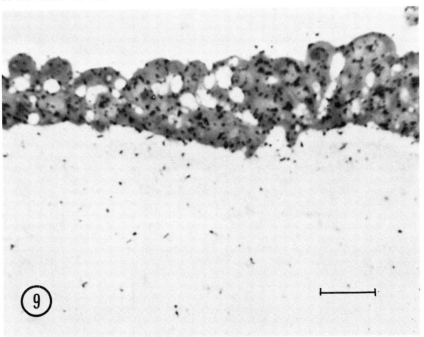

FIG. 9. After a 4-hr incubation with 50 μCi [³H]proline radioautographs show label accumulation over the myocardium and the underlying cardiac jelly of a stage-11 (13-somite) chick embryo, suggesting the production of a proteinaceous material. While these radioautographic data alone are not sufficient, they do suggest that the myocardium may also be producing collagen or a collagenlike molecule. The mark represents 20 μm. Courtesy Dr. Alan Cohen.

velop myofibrils before fusion (Przybylski and Blumberg, 1966; Waterman, 1969) yet seemingly do not divide.

The compatibility of differentiation and cell division in cardiac muscle is certainly not unique. Many other cell types also disobey the "rule" that differentiated cells do not divide. The list now includes smooth muscle (Cobb and Bennett, 1970), pancreatic islet cells (Like and Chick, 1969), and red blood cells (De La Chapelle *et al.*, 1969; Jeter and Cameron, 1971), and it should not be surprising to add cardiac muscle

---

FIG. 8. Carbohydrates are also good extracellular matrix precursors. This radioautograph shows the distribution of label within the heart of a 13-somite chick embryo incubated with 25 μCi [³H]glucosamine for 2 hr. The distribution of glucosamine-derived label is very similar to that obtained with ³⁵SO₄²⁻ (Fig. 4). Again, all the evidence suggests that the myocardium (M) which consists of functional developing myocytes elaborates the embryonic connective tissue. No great concentration of silver grains are seen immediately surrounding the endocardium (E). The mark represents 10 μm.

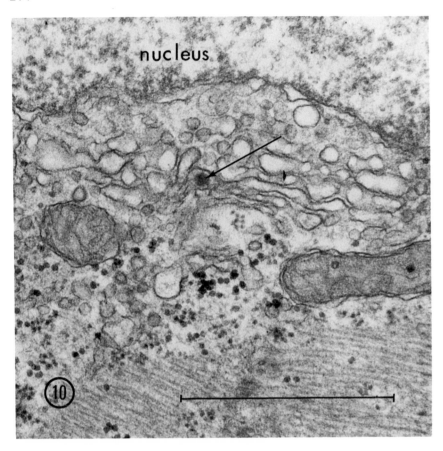

Fig. 10. In this electron micrograph of a portion of a ventricular myocyte from
an 11-day chick embryo, a prominent dense granule is seen within the Golgi
complex (arrow). It has been suggested that such granules represent a secretory
product. The mark represents 1 μm.

to this list. The relationship between mitosis and differentiation for a
number of cell types has recently been discussed by Cameron and Jeter
(1971).

The homogeneity of the early developing myocardial wall also makes
this tissue well-suited for *in vivo* work. For example, it can clearly
be demonstrated by means of radioautography that a variety of precursors to extracellular matrix materials are utilized by both undifferentiated cardiac myoblasts as well as myocytes. There is no indication
that "fibroblasts" are involved in this process. Although the myogenic
regions (presumed to consist of skeletal myoblasts) of the limb bud

TABLE II

DEVELOPMENTAL HISTORY OF CARDIAC AND SKELETAL MUSCLE

		Skeletal muscle	
	Cardiac muscle	Somite	Limb
Precursor tissue organization	Epithelial	Epithelial	Mesenchymal
Developing tissue	Homogeneous epithelium	Homogeneous epithelium	Heterogeneous mesenchyme
Onset of fibrillogenesis	Mononucleated cells	Mononucleated cells	Multinucleated cells (syncytial)
Cell fusion	No	Yes	Yes
Mature form of cells	Discrete cells	Syncytium	Syncytium
Mitosis in undifferentiated cells	Yes	Yes	Yes
Mitosis in differentiated cells	Yes	No	No
Secretion by undifferentiated cells	Yes	?	? (probably yes)
Secretion by differentiated cells	Yes	? (probably no)	? (probably no)

incorporate $^{35}SO_4{}^{2-}$ (Searls, Chapter 9, this volume), no clear-cut evidence for secretion by skeletal myotubes exist.

What is the role, if any, of the matrix produced by cardiac muscle? Does the extracellular cardiac matrix differ in composition at different developmental times? If the mucopolysaccharide constituent is pharmacologically altered, does this change the events of cytodifferentiation of cardiac myoblasts?

There are some tantalizing possibilities. The earlier work of Hauschka and Konigsberg (1966) has shown that skeletal muscle *in vitro* requires collagen for optimal development. More recently Schubert and Jacob (1970) have demonstrated a relationship between the production of extracellular material and differentiation of cultured neuroblastoma cells. It is tempting to propose that cells *in vivo* also require specific sets of environmental conditions with which interactions are mandatory during differentiation. These may not be similar to the classic concepts of an "inducer" molecule, but rather represent complex physical and chemical environments, such as afforded by embryonic extracellular matrix, that trigger a "recognition" mechanism in the developing cells. Since the myocardium is homogeneous and contains only developing muscle cells, these cells may be adapted to regulating their own environment. As has been pointed out, limb skeletal muscle differentiates in an environment containing chondroblasts and true fibroblasts. These true connective tissue cells may have taken over the functions of producing extracellular matrix with which myoblasts interact and which may be extremely important in tissue development. Certainly much of the foregoing is highly speculative, but these possibilities can be tested by experimentation.

How functionally restricted is muscle in the differentiated state? If we ask this question and use cardiac muscle as an example, we see that functional restrictions are not very obvious. Cardiac muscle synthesizes DNA, undergoes mitosis, synthesizes large numbers of myofibrils, and also secretes an extracellular product or products (Manasek, 1970b). If we ask this question using skeletal muscle as a model the answer, as we have seen, is quite different. These differences, explored on a number of levels, have been the central theme of this chapter and it is hoped that further comparative studies will aid in clarifying problems of striated myogenesis.

References

Allen, E. R., and Pepe, F. A. (1965). *Amer. J. Anat.* **116,** 115.

Bloom, S., and Egli, D. (1969). *Proc. Soc. Exp. Biol. Med.* **130,** 1019.

Boyd, J. D. (1960). *In* "Structure and Function of Muscle" (G. G. Bourne, ed.), 1st ed., Vol. 1, pp. 63–85. Academic Press, New York.

Cameron, I. L., and Jeter, J. R., Jr. (1971). *In* "Developmental Aspects of the Cell Cycle" (I. L. Cameron, G. M. Padilla, and A. M. Zimmerman, eds.), pp. 191–222. Academic Press, New York.

Chacko, S. K. (1971). *Anat. Rec.* **169,** 293.

Cobb, J. L. S., and Bennett, T. (1970). *Z. Zellforsch. Mikrosk. Anat.* **108,** 177.

Davis, C. L. (1924). *Anat. Rec.* **27,** 201.

De La Chapelle, A., Fantoni, A., and Marks, P. A. (1969). *Proc. Nat. Acad. Sci. U.S.* **63,** 812.

Fischman, D. A. (1970). *Curr. Top. Develop. Biol.* **5,** 235–280.

Franco-Browder, S., De Rydt, J., and Dorfman, A. (1963). *Proc. Nat. Acad. Sci. U.S.* **49,** 643.

Gaines, L. M. (1960). *Bull. Johns Hopkins Hosp.* **106,** 195.

Gessner, I. H., Lorincz, A. E. and Bostrom, H. (1965). *J. Exp. Zool.* **160,** 291.

Grohmann, D. (1961). *Z. Zellforsch. Mikrosk. Anat.* **55,** 104.

Hauschka, S. D., and Konigsberg, I. R. (1966). *Proc. Nat. Acad. Sci. U.S.* **55,** 119.

Hay, E. D. (1968). *In* "Epithelial-Mesenchymal Interaction" (R. Fleischmajer and R. E. Billingham, eds.), pp. 31–55. Williams & Wilkins, Baltimore, Maryland.

Herrmann, H., Heywood, S. M., and Marchok, A. C. (1970). *Curr. Top. Develop. Biol.* **5,** 181–234.

Holtzer, H., Marshall, J. M., Jr., and Finck, H. (1957). *J. Biophys. Biochem. Cytol.* **3,** 705.

Ishikawa, H., Bischoff, R., and Holtzer, H. (1968). *J. Cell Biol.* **38,** 538.

Jamison, J. D., and Palade, G. E. (1964). *J. Cell Biol.* **23**, 151.

Jeter, J. R., and Cameron, I. L. (1971). *J. Embryol. Exp. Morphol.* **25**, 405.

Johnston, P. M., and Comar, C. L. (1957). *J. Biophys. Biochem. Cytol.* **3**, 231.

Kompmann, M., Paddags, I., and Sandritter, W. (1966). *Arch. Pathol.* **82**, 303.

Konigsberg, I. R. (1963). *Science* **140**, 1273.

Lash, J. W. (1968). *In* "Epithelial-Mesenchymal Interactions" (R. Fleischmajer and R. E. Billingham, eds.). Williams & Wilkins, Baltimore, Maryland.

Like, A. A., and Chick, W. L. (1969). *Science* **163**, 941.

Manasek, F. J. (1968a). *J. Cell Biol.* **37**, 191.

Manasek, F. J. (1968b). *J. Morphol.* **125**, 329.

Manasek, F. J. (1969a). *J. Cell Biol.* **43**, 605.

Manasek, F. J. (1969b). *J. Embryol. Exp. Morphol.* **22**, 233.

Manasek, F. J. (1970a). *Amer. J. Cardiol.* **25**, 149.

Manasek, F. J. (1970b). *J. Exp. Zool.* **174**, 415.

Medoff, J (1967). *Develop. Biol.* **16**, 118.

Olivo, O. M., Laschi, R., and Lucchi, M. L. (1964). *Sperimentale* **114**, 69.

O'Neill, M., and Strohman, R. C. (1969). *J. Cell. Physiol.* **73**, 61.

Ortiz, E. C. (1958). *Arch. Inst. Cardiol. Mex.* **28**, 244.

Orts-Llorca, F., and Gonzáles-Santander, R. (1969) *Rev. Espan. Cardiol.* **22**, 537.

Pager, J. (1968). Doctorate Thesis, Faculty of Sciences, University of Lyon, France.

Parry, W. (1968). *Amer. J. Anat.* **122**, 491.

Przybylski, R. J., and Blumberg, J. M. (1966). *Lab. Invest.* **15**, 836.

Rice, R. V., Moses, J. A., McManus, G. M., Brady, A. C., and Blasik, L. M. (1970). *J. Cell Biol.* **47**, 183.

Rosenbluth, J. (1971). *J. Cell Biol.* **48**, 174.

Rosenquist, G. C., and De Haan, R. L. (1966). *Contrib. Embryol. Carnegie Inst.* **38**, 111.

Rumery, R. E., and Rieke, W. O. (1967). *Anat. Rec.* **158**, 501.

Rumyantsev, P. P., and Snigirevskaya E. S. (1968). *Acta Morphol.* **16**, 271.

Schubert, D., and Jacob, F. (1970). *Proc. Nat. Acad. Sci. U.S.* **67**, 247.

Searls, R. L. (1967). *J. Exp. Zool.* **166**, 39.

Shimada, Y. (1971). *J. Cell Biol.* **48**, 128.

Sissman, N. J. (1966). *Nature (London)* **210**, 504.

Stockdale, F E., and Holtzer, H. (1961). *Exp. Cell Res.* **24**, 508.

Straus, W. L., Jr., and Rawles, M. E. (1953). *Amer. J. Anat.* **92**, 471.

Thorp, F. K., and Dorfman, A. (1967). *Curr. Top. Develop. Biol.* **2**, 151–190.

Waterman, R. E. (1969). *Amer. J. Anat.* **125**, 457.

Weinstein, R. B., and Hay, E. D. (1970). *J. Cell Biol.* **47**, 310.

Weissenfels, N. (1968). *Z. Zellforsch. Mikrosk. Anat.* **86**, 1.

Williams, L. W. (1911). *Amer. J. Anat.* **11**, 55.

Woessner, J. F., Bashey, R. I., and Boucek, R. J. (1967). *Biochim. Biophys. Acta* **140**, 329.

Supplementary References

Since this chapter was written, a number of relevant publications have appeared, some of which are listed below.

Chacko, S. (1973). *Anat. Rec.* **175**, 287.

Doyle, C., Zak, R., and Fischman, D. (1972). *J. Cell Biol.* **55**, 63a.

Fambrough, D., and Rash, J. (1971). *Develop. Biol.* **26,** 55.

Johnson, R. C., Manasek, F. J., Vinson, W. C., and Seyer, J. M. (1973). *Anat. Rec.* **175,** 350.

Lipton, B. H., and Konigsberg, I. R. (1972). *J. Cell Biol.* **53,** 348.

Paterson, B., and Strohman, R. C. (1972). *Develop. Biol.* **29,** 113.

Viragh, S., and Challice, C. E. (1973). *J. Ultrastruct. Res.* **42,** 1.

9

Chondrogenesis

ROBERT SEARLS

In this review I will concentrate on the differentiation of cartilage in the embryonic chick limb. Some of the observations made on chick limb will be compared with observations concerning cartilage differentiation in the chick somite. I will not attempt to give an extensive description of cartilage differentiation in other regions of the chick nor in other organisms.

The differentiation of limb cartilage has been studied for two reasons: (1) Cartilage, muscle, and connective tissue arise during the development of the embryonic chick limb. Cartilage and muscle, in particular, are very different in their cytology, biochemistry, and developmental history. An understanding of the mechanism by which these cell types arise would be very valuable. (2) Cartilage differentiates in the limb in a very definite pattern. The pattern of the cartilages in the chick limb has been the usual criterion for the recognition of proximal or distal limb structures, or of limb polarity, during studies of limb morphogenesis. Interpretation of the many experiments concerning limb morphogenesis would be easier if some understanding of the control of the deposition of cartilage in the limb could be obtained. At the same time, investigations on the differentiation of cartilage in the limb may be guided by techniques developed during previous investigations of limb morphogenesis.

I. Description of Cartilage Differentiation

A. MORPHOLOGICAL AND CYTOLOGICAL CHANGES DURING CARTILAGE DIFFERENTIATION

1. Description of Limb Outgrowth

The limb first becomes visible as an outgrowth at the beginning of stage 17 (Hamburger and Hamilton, 1951). From the beginning of stage 17 to the end of stage 23 the stage of the embryo is determined in large part by the ratio of the length (i.e., craniocaudal distance along the body wall) to the width (proximodistal distance from the body wall to the apex) of the right wing and leg, and for several stages after stage 23 the stage of the embryo is determined mostly by the shape of the wing and leg (Hamburger and Hamilton, 1951). Thus, limbs obtained from embryos of a particular stage are quite uniform in their state of development.

The length of time required for the limb to develop from a barely visible outgrowth to stage 25 has been determined in order that successive differentiative changes may be correlated in time. The length and width of the right wing bud was measured at intervals during embryonic development (Janners and Searls, 1970). The change in width of the wing was plotted against time to give a preliminary indication of the rate of wing outgrowth (Fig. 1). The ratio of length to width (L/W) of the wing that corresponded to each of the width measurements plotted in Fig. 1 was then calculated and L/W was plotted against time (Fig. 2). Since the stage of the chick embryo during this period is defined in part by the ratio of length to width of the right wing bud, the stage of embryonic development from stage 16 to stage 25 could be correlated with time of embryonic development by using Figs. 1 and 2 and by assuming that the limb becomes measurable at about 60 hr of embryonic development.

2. Cytological Changes in the Limb

Cytological changes that may be observed with the light microscope during the differentiation of cartilage in the chick limb were first de-

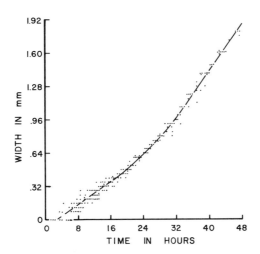

FIG. 1. A standard curve of increase in width of the wing during the first 48 hr of wing outgrowth. The data are measurements made on a total of 43 different embryos. Each wing was measured at least three times during a period of at least 22 hr. Taken from Janners and Searls (1970).

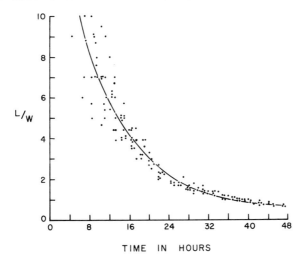

FIG. 2. A standard curve of change in length to width ratio (L/W) of the right wing bud during the first 48 hr of wing development. The L/W for each of the width measurements plotted in Fig. 1 was calculated and plotted at the appropriate time. Taken from Janners and Searls (1970).

scribed by Fell and Canti (1934). These changes have been pictured by Saunders (1948) in his depiction of the stages of limb development and by Jurand (1965). Distinctive cell types as determined by cytological criteria have not been described in the limb mesenchyme from the appearance of the limb at stage 17 to about the middle of stage 22 (Fig. 3). In the middle of stage 22 (when L/W of the right wing is about 1.8) the cells in the proximal half of the limb in the central core begin to become more densely packed and appear to have decreasing amounts of cytoplasm, but no metachromatic material can be detected in the vicinity of these cells (Fig. 4). During stage 24, the mass of condensed mesenchyme becomes Y-shaped when viewed in frontal section. Thus, the future long bones and the position of the prospective joint are demarcated during stage 24. Traces of metachromatic material are found in the cartilage-forming region of the limb starting during stage 25 (Fig. 5). After stage 25, metachromatic material increases in amount until stage 27 when the long bones of the limb are clearly visible in sections as regions of "epitheloid" cells embedded in a lake of metachromatic material.

3. Development of Condensed Mesenchyme in the Limb

The region of the limb in which cartilage will appear, the chondrogenic region, is recognized under the light microscope in part as a region of condensed mesenchyme. The time of appearance of this condensation of mesenchymal cells has been investigated by M. Y. Janners and R. L. Searls (unpublished observations). Wing buds were serially sectioned at 5 μm, a standard field was superimposed on various regions of the sections (Fig. 6) using a drawing tube, and the number of nuclei contained within the standard field was counted. The average number of nuclei counted in the standard field in each region of the wing at stages 19, 22, and 24 is presented in Fig. 7. The average number of nuclei counted in the subridge region seemed to remain almost constant from stage 19 to 24. In contrast, the average number of nuclei counted in the central proximal (chondrogenic) region and in the dorsal and ventral (myogenic) regions of the wing increased about 50% and 25%, respectively.

The number of nuclei counted in the standard field could not be easily translated into the number of cells in a standard volume of tissue. The nuclei are about 4 μm in diameter, so many of the "nuclei" counted in a 5-μm section were portions of nuclei. To calculate the number of cells that would be present in a known volume of tissue from the thickness of the section and the number of nuclei counted, one would have to use various correcting factors. Further, the density of cells in sectioned material varies with the conditions of fixation and embedding experienced by each piece of tissue. However, the number of nuclei counted in the standard field should be directly proportional to the number of nuclei in a standard volume, since the conversion involves multiplication by certain correcting constants. If the number of nuclei in a standard field in a 5μm section is plotted against time of limb development, the slope of the curve obtained should be identical with the slope of the curve that would be obtained if the number of cells in a standard volume could be plotted against time of limb development. To correct for the effects of histological processing, it was assumed that swelling or compression of the tissue was the same throughout each of the fixed and sectioned wings and that the cell density in the subridge region of the wing did not vary during wing development. The number of nuclei in the standard field in the subridge region of the wing was set equal to 100, and the number of nuclei in the standard field in the proximal regions of each wing was calculated as a fraction of the number in the subridge region. The values obtained were then plotted against time of wing outgrowth, ratio of length to width of the wing, and the stage of the embryo (Fig. 8).

The density of cell packing was about the same throughout the wing until about stage 20 (about 20 hr of wing development). The density of cell packing then began to increase in all proximal regions of the wing. At stage 24 (about 38 hr of wing development), the cell density in the central proximal (chondrogenic) region of the wing was 1.5 times the density at stage 19, and the cell density in the dorsal and ventral proximal (myogenic) regions of the wing was 1.2 times the density at stage 19. The density of cell packing did not appear to change further during stages 25 and 26.

4. Ultrastructural Changes in the Limb

Ultrastructural changes that occur during the differentiation of cartilage in the chick limb have been described by Goel (1970) and in the mouse limb by Godman and Porter (1960). Although the results obtained in these studies are very similar, Goel gives the stages in embryonic development more clearly than Godman and Porter so I will concentrate on his observations.

The earliest cells examined in both of these studies are called "mesenchyme." They are reported by Goel to be from the limb of a stage-23 embryo and described by Godman and Porter as from an area of "massed mesenchyme which constitute the centers of chondrification." It is clear that these are cells of the chondrogenic region and that neither of the papers examines possible ultrastructural changes that may occur during the formation of the region. Goel states that these "mesenchyme" cells "appear from the unspecialized arrangement of the cell organelles to be undifferentiated." Next, Goel describes "prechondrogenic" cells from

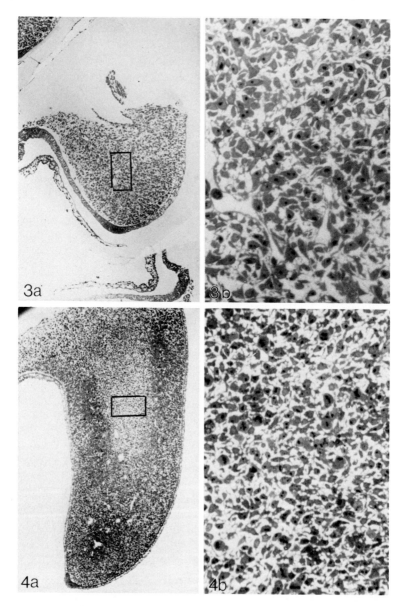

FIGS. 3–5. Wing buds fixed in glutaraldehyde buffered with phosphate to pH 7.8 (Coleman *et al.*, 1969), embedded in Araldite (Cargille), sectioned at 1.0 μm, and stained with methylene blue–Azure II.

FIG. 3a. Stage-19 wing bud photographed at 100×.

FIG. 3b. The indicated area in Fig. 3a photographed at 400×.

FIG. 4a. Stage-24 wing bud photographed at 100×.

FIG. 4b. The indicated area in Fig. 4a photographed at 400×.

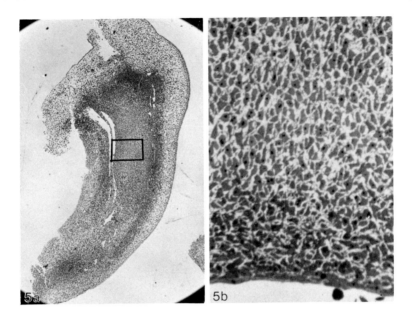

Fig. 5a. Stage-26 wing bud photographed at 60×.
Fig. 5b. The indicated area in Fig. 5a photographed at 400×.

the limb of a stage-26 embryo. These are said not to differ in their ultrastructure from the undifferentiated mesenchyme cells. Reportedly, the extracellular spaces surrounding these "prechondrogenic" cells are electron translucent.

The next cells described by Goel are "chondrogenic" cells from the limb of a stage-31 embryo. These cells are described as having a well-developed endoplasmic reticulum, a well-developed Golgi apparatus,

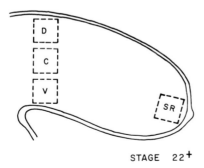

STAGE 22⁺

Fig. 6. A diagram illustrating the position of the fields counted: SR, the subridge region; D, the dorsal proximal region; C, the central proximal region; and V, the ventral proximal region. Taken from Janners and Searls (1970).

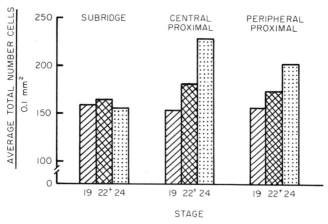

Fɪɢ. 7. The average cell density (average number of cells in a 0.1 mm² field) in the subridge, central proximal and peripheral (dorsal and ventral) proximal regions of the wing at stage 19, stage 22⁺, and stage 24. Data from M. Y. Janners. (unpublished observations).

and a "scalloped or bayed appearance, probably due to the opening out of the vacuoles containing extracellular material." These cells are very similar to the "early chondroblasts" and "developed chondroblasts" described by Godman and Porter. At stage 31, the matrix of the cartilage is described by Goel as an "amorphous, largely electron-translucent ground substance with a fair number of short unbanded fibers embedded in it. At times some electron-dense granules are also observed on the fibers."

Thus, the cytological changes that have been observed with the light microscope (the appearance of the chondrogenic region during stage 22 and the appearance of extracellular metachromatic material in the chondrogenic region during stage 25) do not seem to be accompanied by ultrastructural changes easily detected in the electron microscope.

5. *Cytological and Ultrastructural Changes in the Somite*

The cytological changes that occur in the somite between stage 11 and 18 have been described by Hay (1968) at both light microscope and the electron microscope levels. Changes that occur in the sclerotome between stage 18 and 26 have been observed with the light microscope (Kvist and Finnegan, 1970a) and to some degree with the electron microscope (O'Connell and Low, 1970).

The somite forms from "primary mesenchyme" as a sphere of epithelial cells. Later, the cells in the ventromedial wall of this sphere of epithelial cells disperse forming a "secondary mesenchyme." The "secondary

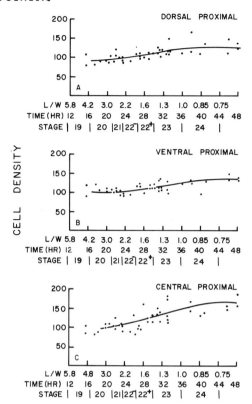

Fig. 8. Plots illustrating the increase in cell density in (A) the dorsal proximal region, (B) the ventral proximal region, and (C) the central proximal region. The cell density is expressed as a percentage of the density in the subridge area when the number of cells in a 0.1 mm² field in the subridge area is set equal to 100. The data are presented in this way to correct for the variable effects of conditions of fixation, embedding, and sectioning in different wings. Data from M. Y. Janners. (unpublished observations).

mesenchyme" cells migrate ventrally and medially until they come to surround closely the neural tube and the notochord. These cells then make up the sclerotome of the somite (Hay, 1968).

Extracellular metachromatic material appears in the sclertome in the vicinity of the notochord between stage 22 and 26, probably during stage 24 (Kvist and Finnegan, 1970a). O'Connell and Low (1970) have investigated extracellular microfibrils near the notochord between stage 7 and 29, but did not describe any ultrastructural changes in the sclerotome cells in the vicinity of the notochord during this period of development.

B. Changes in Rate of Cell Division in the Wing

Changes in rate of cell division in various regions of the wing (Fig. 6) during development have been measured by determining the labeling index, i.e., the percentage of cells that become labeled during a short period of tritiated thymidine uptake (Fig. 9) (Janners and Searls, 1970). The labeling index in all proximal regions of the wing is constant until about stage 21 (about 22 hr of wing development) when the labeling index begins to decrease. In the proximal dorsal and ventral (myogenic) regions the labeling index decreases until stage 23 (about 34 hr of wing development). The labeling index in the proximal dorsal and ventral (myogenic) regions does not change further between stage 23 and 26. In the proximal central (chondrogenic) region of the wing, the labeling index decreases in parallel with the decrease in labeling index in the myogenic regions until the middle of stage 22 (from about 22 hr to about 26 hr of wing development). During stage 22, when the L/W of the right wing is about 1.8, the labeling index in the central proximal (chondrogenic) region begins to fall more rapidly than the labeling index in the myogenic regions. The labeling index in the central proximal region stops decreasing during stage 23 (about 34 hr of wing development) and does not change further between stage 23 and 26.

Continuous labeling experiments were done at stage 19 and at stage 24 in order to discover the cause of the decrease in labeling index. The data obtained from the continous labeling experiments were treated according to the procedure of Okada (1967) to obtain cell division parameters (Tables I and II). The decrease in labeling index in the proximal dorsal and ventral (myogenic) regions was found to result from the withdrawal of approximately 25% of these cells from the division cycle.

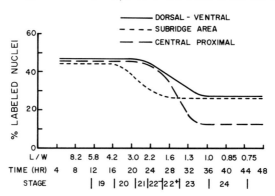

Fig. 9. A comparison of the changes in labeling index in the various regions of the wing during the first 48 hr of wing development. Taken from Janners and Searls (1970).

TABLE I

Duration[a] of the Cell Cycle[b] and Its Component Phases at Stage 19

Region	$G_2 + M$	G_1	$G_1 + G_2 + M$	S	T	PI[c]
SR	2.8	5.1	7.9	5.6	13.5	1.00
C	2.5	4.0	6.5	5.6	12.1	1.00
D	2.5	4.0	6.5	5.6	12.1	1.00
V	2.5	4.0	6.5	5.6	12.1	1.00

[a] Duration is given in hours.
[b] T, cell cycle.
[c] PI is the proliferative index or the fraction of the population which is actively dividing.

TABLE II

Duration[a] of the Cell Cycle[b] and Its Component Phases at Stage 24

Region	$G_2 + M$	G_1	$G_1 + G_2 + M$	S	T	PI[c]
SR	2.5	5.1	7.6	5.6	13.2	0.75
C	1.5	6.1	7.6	5.6	13.2	0.25
D	2.0	5.6	7.6	5.6	13.2	0.75
V	2.0	5.6	7.6	5.6	13.2	0.75

[a] Duration is given in hours.
[b] T, cell cycle.
[c] PI is the proliferative index or the fraction of the population which is actively dividing.

In the proximal central (chondrogenic) region the decrease was found to result from the withdrawal from the division cycle of 75% of the cells in that region.

Since the decrease in labeling index in each case is caused by a decrease in the proliferative index, the first nondividing cells must appear in the proximal regions during stage 21. The decrease in labeling index occurs over a period of about 12 hr, or about the duration of one division cycle. Since the labeling index does not change after stage 23, cells must continue to withdraw from the division cycle at a slow rate during succeeding stages of limb development.

C. Biochemical Changes during Cartilage Differentiation

1. Mucopolysaccharide Synthesis in the Limb

a. Extraction of Mucopolysaccharide from the Precartilaginous Limb. Mucopolysaccharides have been extracted from limbs of the embryonic chick at stages before extracellular metachromatic material can be detected in the cartilage-forming region. Material that contains ester sulfate, is hyaluronidase sensitive, and that migrates in an electrophoretic field with chondroitin sulfates A and/or C has been isolated from the limb-forming region of stage-15 embryos and from the limbs of stage-16 to stage-24 embryos (Franco-Browder *et al.*, 1963; Medoff, 1967). A material that contains ester sulfate, migrates just behind chondroitin sulfate A and/or C, and that is hyaluronidase insensitive has been extracted from limbs of embryos at stage 19 or younger (Medoff, 1967).

b. Distribution of Mucopolysaccharide Synthesis in the Limb. The distribution of mucopolysaccharide-synthesizing activity within the limb at various times during limb development *in ovo* has been determined using ^{35}S and autoradiography (Searls, 1965a). Eggs containing embryos ranging from stage 15 to 27 were administered the same dose of [^{35}S]sulfate (about 50 μCi, the exact doses varied from experiment to experiment but was constant within a single experiment). The limbs of these embryos were fixed 2 hr after administration of the [^{35}S]sulfate and sectioned. The sections from all of the limbs from a single experiment were dipped in photographic emulsion and developed at the same time. The autoradiographs were examined at low power under dark-field illumination to obtain a gross view of the distribution of silver grains over the sections, and thus provided a picture of the distribution of mucopolysaccharide-synthesizing activity.

No difference in uptake of [^{35}S]sulfate was found between the flank regions and the limb regions at stages 16 and 17; [^{35}S]sulfate appeared to be taken up uniformly in both regions, and when the density of silver grains was compared it seemed to be taken up at the same rate in both regions. Uptake of [^{35}S]sulfate in the flank was not examined after stage 17. Between stage 17 and the middle of stage 22, [^{35}S]sulfate was taken up throughout the limb, no regions of higher or lower uptake could be detected in cross or frontal section.

A differential uptake of [^{35}S]sulfate first became evident in the proximal half of the limb during stage 22, when the L/W of the wing was about 1.8. At that time, the rate of sulfate uptake began to increase in the central core and to decrease on the periphery. The contrast between the central and peripheral rates of uptake became more pronounced between stage 22 and 24 so that by stage 25 an exposure of the autoradiographs that gave a high concentration of silver grains over the cartilage-forming region gave a concentration of silver grains over the prospective soft tissue regions not in excess of background. However, uptake of [^{35}S]sulfate continues in the soft tissue regions during stage

25 and 26; a sufficiently long exposure of the autoradiographs will show silver grains over the soft tissue regions well in excess of background.

During the development of this pattern of sulfate uptake, a gradient of uptake is never observed. The uptake within a particular region of the limb appears to be uniform, and a sharp boundary is observed between one region and a neighboring region. Within a region, one does not observe "hot spots" that would suggest that some of the cells are capable of more rapid uptake than neighboring cells.

These observations were made by administering the same amount of [^{35}S]sulfate to embryos of various stages and allowing the same period for uptake. The limbs were fixed, sectioned, dipped in photographic emulsion, and developed at the same time. However, the results may not be interpreted quantitatively since the size of the pool of cold sulfate may change during the development of the embryo.

c. Identification of Radioactive Material Observed in the Autoradiographs. All of the radioactivity detected in these autoradiographs was due to mucopolysaccharide (Searls, 1965b). Limbs from embryos that had taken up [^{35}S]sulfate during stage 22 and 23 were fixed and washed as for autoradiography, and the radioactivity remaining in the limbs was then extracted. After digestion with 0.3 M NaOH, digestion with papain, and fractionation of the Ca salts of the radioactive material with ethanol, the radioactivity (73% recovery) was found to be excluded by a Sephadex G-50 column. After treatment with hyaluronidase, approximately 90% of the radioactivity was retarded on the same Sephadex G-50 column. Based on the solubility properties of the Ca salts of the radioactive material and the hyaluronidase sensitivity of the various fractions, it was suggested that limbs from stage-22 to stage-23 embryos contain about 60% chondroitin sulfate A, 10% chondroitin sulfate B, and 30% chondroitin sulfate C. This is very similar to the composition of the mucopolysaccharides of limb and sternal cartilages from 17-day embryos (60% chondroitin sulfate A, 30% chondroitin sulfate C, and 10% unsulfated chondroitin) (Mathews, 1967; Saito *et al.*, 1968), except that mature cartilage contains no chondroitin sulfate B. The chondroitin sulfate B may have come from some other region of the limb, perhaps the basement membrane, since complete limbs from stage-22 to stage-23 embryos were extracted.

d. Investigation of Heterogeneity in the Mesenchyme of the Early Limb. Experiments were done to discover whether the uptake of [^{35}S]sulfate into mucopolysaccharide observed prior to stage 25 was due solely to prospective cartilage (or connective tissue) cells present in the limb mesenchyme (Searls, 1967). Prior to the middle of stage 22, uptake of [^{35}S]sulfate into mucopolysaccharide is uniform throughout the limb mesenchyme. If this uptake is caused by cells destined to be cartilage cells, then either (a) the limb mesenchyme prior to the middle of stage 22 must be a random mixture of prospective cartilage cells and prospective soft tissue cells so that each nonsynthesizing cell is surrounded by synthesizing cells and the distribution of synthetic activity appears to be homogeneous, or (b) all of the cells must be prospective cartilage cells (or at least fibroblasts) and the muscle cells must migrate in from outside the limb.

Limb buds were obtained from stage-21 or early stage-22 embryos that had been uniformly labeled with tritiated thymidine, the limb mesoblasts were isolated surgically, and blocks of labeled mesoderm were implanted into the wings of unlabeled embryos (stage 20 to 24) that were growing *in ovo*. The host wings were fixed 12 to 72 hr after the operation and examined by autoradiography to determine the fate of the implanted cells. Similar experiments were done in which the mucopolysaccharide-synthesizing cells in the limb mesoderm were labeled specifically with [^{35}S]sulfate and the host limbs were fixed 8 to 120 hr after the operation.

Irrespective of the label used ([^{3}H]thymidine or [^{35}S]sulfate) or the age of the host (stage 20 to 24), the labeled cells were always found in a single cohesive mass. The implanted cells were sometimes found entirely in the soft tissue region of the host wing, sometimes in both the soft tissue region and the cartilage region of the host wing, and sometimes almost entirely in the cartilage region of the host wing. Wherever the cells were placed, they were integrated into the host wing pattern, so that they could only be recognized by presence of the label. The host limbs were fixed at intervals so that any movement of the implanted cells could be detected if they moved within the host limb and then reaggregated. No evidence of cell movement in or out of the block of implanted cells during development of the host wing was obtained. [^{35}S]sulfate that had been incorporated into mucopolysaccharide by limb mesenchyme cells prior to stage 22 or by cells of the chondrogenic region prior to stage 23 could be detected in the muscle regions as well as in cartilage and connective tissue regions 5 days after the operation.

Thus, the ability to synthesize mucopolysaccharide seems to be a property of all limb mesenchyme cells. Those cells in the early limb that through division give rise to cells in the central proximal region differentiate into cartilage cells, cells that synthesize mucopolysaccharide at an enhanced rate. The limb mesenchyme cells in the early limb that through division give rise to cells in the dorsal and ventral proximal regions differentiate into soft tissue cells, cells that synthesize mucopoly-saccharide at a reduced rate.

e. *Mucopolysaccharide Synthesizing Enzymes in the Limb.* Variation during limb development in the activity of some of the enzymes involved in the synthesis of mucopolysaccharide has been studied by Medoff (1967). Four enzymes or enzyme systems were studied: the sulfate-activating enzymes, uridine diphosphate-D-glucose dehydrogenase, uridine diphosphate-N-acetylglucosamine-4-epimerase, and the mucopolysaccha-ride-polymerizing system. The changes in activity of the sulfate-activating enzymes, uridine diphosphate-D-glucose dehydrogenase, and uridine diphosphate-N-acetylglucosamine-4-epimerase that were observed during limb development *in vivo* are shown in Fig. 10. The polymerizing enzymes were demonstrated in the mesoderm of the limbs of stage-20 embryos, but changes in this activity during limb development were not measured.

The sulfate-activating system was found to increase in activity about fourfold between stage 19 and 5 days of embryonic development (stage 25?). Uridine diphosphate-D-glucose dehydrogenase and uridine diphosphate-N-acetylglucosamine-4-epimerase each increase about 2.5-fold between stage 19 to 20 and 6 days of development (stage 27 to 28). Since whole limbs were used as a source of enzyme, it is obvious that the increase in activity would be much greater if only the cartilage-forming cells had been assayed for activity. If it is assumed that

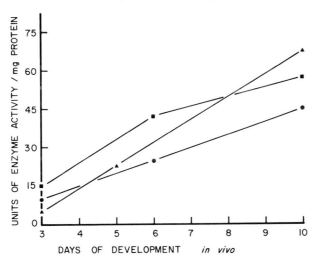

FIG. 10. Developmental kinetics of sulfate-activating enzymes (▲——▲), UDPG-dehydrogenase (●——●), and UDPGNAc-4-epimerase (■——■) in fresh whole limbs. Modified from Medoff (1967).

all of the cells in the limb synthesize mucopolysaccharide at stage 19 to 20, that at 5 to 6 days of embryonic development the limb contains 10% to 20% cartilage-forming cells, and that only the cartilage-forming cells synthesize mucopolysaccharide at 5 to 6 days of embryonic development, then between stage 19 to 20 and 5 to 6 days of embryonic development the sulfate-activating system may increase in activity by twenty to fortyfold in the cartilage-forming cells and the other enzymes may increase in activity by ten- to twenty-fivefold.

In vivo changes in rate of mucopolysaccharide synthesis in the cartilage-forming region and in the prospective soft tissue regions of the limb have not been determined. The changes in enzyme activity during limb development that have been reported by Medoff appear, however, to be of the right order of magnitude to explain the changes in sulfate uptake observed in autoradiographs.

2. *Biochemical Changes in the Somite*

Johnson and Comar (1957) published a study of the uptake of [^{35}S]sulfate by axial tissues in which they injected radioactive sulfate at 0 hr of incubation and fixed at various times thereafter. The distribution of radioactivity in the embryos was determined by autoradiography. This differs from all of the more recent studies in which the embryo has been allowed to accumulate radioactive sulfate for only a few hours. Thus, Johnston and Comar could not determine changes in the pattern of mucopolysaccharide synthesis but only changes in the pattern of mucopolysaccharide deposition. At the earliest stage examined, stage 3, they found an extensive but uniform deposition of [^{35}S]sulfate. By stage 18, they observed deposition of [^{35}S]sulfate in the vicinity of the notochord and in the sclerotome of the somite.

Autoradiographs prepared from $2\frac{1}{2}$-day embryos after a short period of [^{35}S]sulfate uptake show radioactivity in the sclerotome of the somite as well as in many other tissues of the embryo (Lash, 1968a). A material that contains ester sulfate and that cochromatographs with chondroitin sulfate can be extracted from the somites, mesonephros, endoderm, and ectoderm of $2\frac{1}{2}$-day embryos (Lash, 1968b). Uncultured somites from $2\frac{1}{2}$-day embryos that were isolated without the use of trypsin have been demonstrated to have the sulfate-activating enzymes and the enzymes that convert glucosamine to uridine diphospho-N-acetylgalactosamine.

Kvist and Finnegan (1970b) surgically isolated the axial regions of chick embryos (including the sclerotome, myotome, dermatome, and epidermis but excluding all tissues ventral to the notochord including the limb buds) and extracted these for glycosaminoglycans. The axial tissues were washed, defatted, and digested with papain and trypsin. After re-

TABLE III

FRACTIONATION OF MUCOPOLYSACCHARIDE ISOLATED FROM THE AXIAL TISSUES
OF CHICK EMBRYOS OF VARIOUS STAGES[a]

Stage	Uronic acid/gm dry weight of tissue (μmoles)	μmoles Uronic acid/gm dry weight			Total μmoles recovered
		Fraction 1	Fraction 2	Fraction 3	
17/18	7.62	4.02	1.61	0.16	5.79
21/22[b]	10.71	4.98	3.35	0.07	8.40
24/25	10.21	5.05	3.63	0.16	8.84
26[b]	12.01	4.24	5.05	0.26	9.56
27/28[b]	16.40	4.19	10.08	0.94	15.21

[a] Taken from Kvist and Finnegan, 1970b.

[b] The values given for stage 21/22, stage 26, and stage 27/28 are averages of the several values given at those stages by Kvist and Finnegan.

moval of residual protein with 10% TCA, the glycosaminoglycans were fractionated by column chromatography. The results are given in Table III. Stage 17 to 18 chick embryo axial tissues contained 7.62 μmoles of uronic acid per gram dry weight. By stage 27 to 28, this amount had little more than doubled to 16.40 μmoles of uronic acid per gram dry weight. However, at stage 17 to 18 the first fraction off the column was 69% of the total while the second fraction was 28% of the total. At stage 27 to 28, the first fraction off the column had not increased in amount and was only 28% of the total while the second fraction off the column had increased greatly and was 66% of the total. The first fraction off the column contained glucosamine and glucuronic acid and was identified by paper electrophoresis as hyaluronic acid. The second fraction off the column contained galactosamine and glucuronic acid with faint traces of glucosamine and perhaps L-iduronic acid. On paper electrophoresis, the second fraction off the column moved behind chondroitin-4-sulfate and approximately even with dermatan sulfate, but the solubility of the calcium salts of this material in ethanol indicated that it was not dermatan sulfate. Its constituents suggest that it must be chondroitin sulfate A and/or C.

D. COLLAGEN

The presence of collagen in the limb buds of stage-17 to stage-24 chick embryos could not be demonstrated histochemically using Gomori's Trichrome stain, and no hydroxyproline could be detected biochemically in limb buds from those stages (Mottet, 1967). Traces of hydroxyproline could be detected in 5-day limb buds (stage 26 to 27; about 0.05 μg hydroxyproline per milligram wet weight). Collagen was not stainable in the limb bud as late as stage 26. During stage 27, Gomori's Trichrome stain revealed traces of collagen at the ectoderm–mesoderm interface, and in precartilage areas. Too few studies of the limb at the ultrasturctural level have been published for statements to be made concerning the appearance of collagenlike filaments in the cartilage-forming region of the limb. The presence of collagen bundles at the ectoderm–mesoderm interface under the apical ectodermal ridge in 3- to 4-day wing buds (stage 19 to 22?) has been reported (Saunders et al., 1964).

Several studies have been made of collagen synthesis in axial tissues. Deuchar (1963) has observed that higher concentrations of radioactivity from injected proline accumulate in the vicinity of the chorda starting from about 3 days of incubation. More significantly perhaps, O'Connell and Low (1970) have observed extracellular microfibrils between the head process and the medullary plate as early as stage 7, many extracellular microfibrils in the vicinity of the notochord by stage 16, and extracellular microfibrils extending from the notochord into the somatopleure in the region of the notochord by stage 20. They suggest that these microfibrils may represent primitive collagen, or collagen precursor material. If this suggestion is correct, then collagenlike molecules are present in the axis of the embryo long before the appearance of metachromatic material in that region.

E. STABILIZATION OF CARTILAGE PROPERTIES IN THE LIMB

When a block of limb mesenchyme from the limb of a stage-22 to stage-24 embryo was implanted into an ectopic site in a host wing, the block of mesenchyme was incorporated into the host wing pattern. Those cells in the implanted block that came to lie in the cartilage region of the host wing formed cartilage and those cells that came to lie in the soft tissue region of the host wing formed soft tissue. The implanted block of tissue did not form ectopic cartilage (cf. Section I,C,1,c).

Experiments were done to determine when cartilage properties become stabilized in the cartilage-forming mesenchyme of the limb (Searls and Janners, 1969). The particular definition of "stabilized" used here is the ability to form cartilage in the soft tissue region of a host limb. The cartilage-forming region was isolated from the limbs of stage-23 to stage-27 embryos that had been labeled with [³H]thymidine. Blocks of mesoderm from the cartilage-forming regions were implanted into ectopic sites in the wings of unlabeled embryos that were stage 23 to stage 27. After 48 hr of further development, the host wings were examined to determine whether the implanted block of tissue had formed ectopic cartilage.

The results of these experiments are in Table IV. Cells from the cartilage-forming region of the limbs from embryos stage 22 to 24 rarely formed ectopic cartilage when implanted in host limbs from embryos stage 22 to 24. Cells from the cartilage-forming region of limbs from embryos stage 25 to 27 almost always formed ectopic cartilage when implanted in host limbs of stage-22 to stage-24 embryos. The ectopic cartilage observed within blocks of cells from the cartilage-forming region of stage-25 embryos that had been implanted into wings of stage-22 to stage-24 embryos was true cartilage in the sense that mucopolysaccharide was being synthesized in the ectopic cartilage 48 hr after the operation. Thus, cartilage properties become stabilized (by the definition given above) between stage 24 and 25, precisely at that time when they begin to be surrounded by small amounts of extracellular metachromatic material.

Cells from the cartilage-forming region of limbs from embryos stage 23 and 24 produced ectopic cartilage with increased frequency when implanted in older host limbs, suggesting that the control exerted by older host limbs is less stringent and that the approach to "stability" is progressive.

When the stability of cartilage properties in the cartilage-forming cells is defined as the ability to form cartilage when grafted to some region of the embryo other than the limb, or when placed in organ culture,

TABLE IV

A SUMMARY OF THE NUMBER OF OPERATIONS AT EACH STAGE AND THE NUMBER THAT FORMED ECTOPIC CARTILAGE[a]

Donor stage	Host stage 22, 23, 24			Host stage 25, 26, 27		
	Total number	Ectopic cartilage	Ectopic cartilage (%)	Total number	Ectopic cartilage	Ectopic cartilage (%)
22	26	0	0	10	0	0
23	21	2	10	15	4	27
24	23	3	13	18	15	83
25	16	15	94	17	15	88
26/27	16	15	94	29	24	83

[a] Data taken from Searls and Janners, 1969.

it is found that a limb mesoblast will always form cartilage (Zwilling, 1964). However, this observation may not be useful since cells from the prospective soft tissue region of a stage-24 limb will form cartilage in organ culture at the same rate and with the same frequency as cells from the prospective cartilage region of a stage-24 limb (Zwilling, 1966).

F. Summary of Observed Changes

At the initiation of limb outgrowth, the cells in the limb appear to be identical cytologically (Section I,A,2), to be incorporating [^{35}S]sulfate at a uniform rate (Section I,C,1), to be dividing at a uniform rate (Section I,B), and to be packed at a constant density (Section I,A,3). Limb mesenchyme cells differentiate as cartilage or as soft tissue according to their position in the limb; when they are placed in a particular position in the limb, they differentiate in a manner appropriate to their new position (Section I,C,4). No evidence exists that would suggest that the cells in the early limb do not form a homogeneous population of "limb mesenchyme" cells.

The first change observed in the proximal regions of the limb is an increase in cell density starting at about stage 20 (Section I,A,3). The proliferative index begins to decrease (recognized by a decrease in labeling index) in the proximal regions of the wing during stage 21 (Section I,B). However, the central proximal region (the chondrogenic region) first becomes different from the dorsal and ventral proximal regions (the myogenic regions) during stage 22, when the L/W of the right wing is 1.8 (Section I,B). At that time, the labeling index in the chondrogenic region begins to be clearly lower than the labeling index in the myogenic region, the rate of sulfate uptake in the chondrogenic region becomes clearly greater than in the myogenic regions, and in the light microscope the cells in the chondrogenic region begin to appear different from the cells in the myogenic regions.

It is difficult to conclude that any of these changes is responsible for any other. The increase in cell density probably does not lead to the decrease in labeling index. The labeling index in the subridge region of the limb begins to decrease sooner than in the proximal region, at about stage 20 (Section I,B), but there does not appear to be a change in the cell density in that region (Section I,A,3). The labeling index begins to decrease in all of the proximal regions at the same time; whatever causes the decrease in labeling index in the proximal regions must act throughout these areas and cannot be responsible for delimiting the chondrogenic region from the myogenic regions. The chondrogenic region first becomes different from the myogenic region by each of the several criteria that have been investigated at precisely the same time, during stage 22 when the L/W of the right wing is 1.8. One cannot argue that any of these leads to another, but rather all of them must result from some as yet unknown influence.

After the middle of stage 22, the cells change in appearance slowly and continuously so that by stage 31 they have the appearance of typical cartilage cells. The rate of mucopolysaccharide synthesis changes slowly and continuously so that by stage 25 essentially all of the sulfate uptake is in the cartilage-forming region. In contrast, the changes in proliferative index are complete by stage 23; there is no further change between stage 23 and 26; and the changes in cell density are complete by stage 24.

The major and "discontinuous" change that occurs after the middle of stage 22 is the appearance of extracellular metachromatic matrix at stage 25 (Section I,A,2) and the simultaneous stabilization of the cells as cartilage cells (Section I,D). The appearance of metachromatic extra-cellular matrix at stage 25 might be considered simply a quantitative phenomenon. The cells are synthesizing mucopolysaccharide at an en-hanced rate, and the mucopolysaccharide is not being diluted by rapid cell division. Thus, the mucopolysaccharide surrounding each cell in-creases in amount until it can be visualized with stains. The stabilization of cartilage properties might also be considered a quantitative phenome-non. If a block of cells is placed in a host limb of a stage-24 or younger embryo, the cells of the cartilage-forming region clearly become stabilized between stage 24 and 25 (Section I,D, Table III). However, if the host limb is stage 25 or older the stabilization appears to be progressive. Cells from stage-22 limbs regulate completely, cells from the cartilage-forming region of stage-23 limbs regulate 73% of the time, and cells from the cartilage-forming region of stage-24 limbs regulate only 17% of the time. Stabilization appears to result from the accumula-tion of some "stabilizing material," perhaps mucopolysaccharide or mucopolysaccharide-synthesizing enzymes. It is clear, however, that it does not result from the decrease in proliferative index or from the increase in cell density. A block of cells from the cartilage-forming region of a stage-24 limb in which the cells have reached a maximum density and in which 75% of the cells have ceased dividing is able to regulate completely when implanted into a stage-24 host limb. A block of cells from the cartilage-forming region of a stage-25 limb in which the cells have not changed in density from stage 24 and in which the cells have not changed in proliferative index from stage 24 is not able to regulate in a similar host limb.

The changes observed during the differentiation of cartilage in the somite appear to be quite different from those observed in the limb. The somite is formed by the aggregation of the cells of a "primary mesenchyme" into an epithelial sphere. Later, the cells at the ventral medial wall of this epithelial sphere disperse and migrate ventrally and medially as a "secondary mesenchyme." This secondary mesenchyme forms the sclerotome of the somite. The conversion of an epithelium into a mesenchyme would appear to be the step in the somite analogous to the formation of the chondrogenic region in the limb. However, the steps in cartilage differentiation following the formation of the sclerotome of the somite do not seem to follow directly, slowly, and continuously from the formation of the sclerotome as do the steps in cartilage differentiation in the limb; a later interaction between the cells of the sclerotome and the notochord or neural tube appears to be essential for cartilage differentiation in the somite (Holtzer and Detwiler, 1953; Watterson *et al.*, 1954; Strudel, 1953).

II. Control of Cartilage Differentiation

The control of cartilage differentiation in the somite has been reviewed recently (Lash, 1968a,b; Holtzer, 1968; Holtzer and Matheson, 1970) and will not be considered here. This discussion will be limited to the control of cartilage differentiation in the limb.

A. Mosaic Development

The earliest research on morphogenesis of the embryonic chick limb suggested that it is a mosaic system (Huxley and de Beer, 1934). Maps of the early wing bud giving the location of the cells that will form each of the cartilaginous wing structures have been prepared by Saunders (1948). Similar maps for the leg have been prepared by Hampe (1959). When a leg bud was divided into halves or thirds by cutting in either the craniocaudal or in the proximodistal plane and the halves or thirds were allowed to develop either *in situ* or as chorioallantoic or intracoelomic grafts the portions generally gave rise only to those cartilaginous limb structures that would be predicted from the appropriate map (Murray and Huxley, 1925; Murray, 1926; Selby and Murray, 1928; Saunders, 1948). The tip of a limb bud grafted to the dorsal surface of the limb, to the somites, to the chorioallantoic membrane, or placed in avascular culture will produce only distal cartilaginous limb structures (Amprino and Camosso, 1959, 1966; Searls, 1968). [However, a distal sliver consisting of the complete distal border of a stage-19 limb bud developed into a fairly good limb with all limb elements represented when grafted to the flank (Zwilling, 1956).] When a complete stage-20 to -21 limb bud was grafted on to a stage-20 to -21 limb bud from which the tip had been removed, the composite limb frequently gave rise to all of those cartilaginous limb structures that would be expected from both the base and the graft. That is, if the base was wing and the graft was wing, the composite gave rise to a sequence of humerus, radius and ulna, humerus, radius and ulna, wrist and hand (Kieny, 1964a,b). [However, if a complete stage-19 limb bud was grafted to a stage-22 limb bud from which the tip had been removed, the result frequently was normal limb development (Hampe, 1959).] When a particular region of a limb, such as the prospective radius and ulna, was cut from a limb and grafted to a foreign site, the graft usually gave rise to those cartilaginous limb structures expected from that region of the limb (Hampe, 1959; Kieny, 1964a). These observations suggest that the cartilaginous structures of the limb are at least implicit in the limb at a very early stage of development.

In contrast, a number of experiments indicate that the limb mesenchyme is completely regulative. (1) If an apical ectodermal ridge was grafted to any area of a limb mesoblast, even at the base of the limb, the mesoderm under the grafted ridge grew out to produce distal cartilaginous limb structures (Zwilling, 1956; Saunders and Gasseling, 1968). (2) If a portion of the preaxial border of a limb is grafted to the dorsal surface of the limb, the grafted portion will grow out to produce distal cartilaginous limb structures (Amprino and Camosso, 1965). The preaxial border of the limb would normally form only dermis and perhaps some muscle. (3) If cells from a proximal region of the limb are grafted directly under the apical ectodermal ridge, these cells form distal cartilaginous structures. This is true even if the cells are obtained from the prospective soft tissue region of a stage-24 limb (Saunders et al., 1957, 1959; Cairns, 1965). (4) If portions of a limb mesoblast, or disaggregated and completely mixed limb mesoderm cells, are placed in a limb bud ectodermal jacket and grafted to the dorsal surface of the limb or to the chorioallantoic membrane, the graft will form cartilaginous limb structures (Zwilling, 1964). (5) If a portion of the limb is cut away leaving an intact apical ectodermal ridge, normal development ensues (Hampe, 1959; Kieny, 1964a). Normal development may be obtained even if 90% of the limb mesoderm is removed (Barasa, 1964). (6) If a block of limb mesoderm from any region of a limb is placed in organ culture or grafted to the chorioallantoic membrane, the block of tissue will form no more than a nodule of cartilage (Zwilling, 1964). A block of tissue from the chondrogenic region of a stage-24 limb will form cartilage no more quickly and in no greater amount than a block of tissue from the myogenic region (Zwilling, 1966). (7) If a block of limb mesoderm from any region of a limb prior to stage 22 or from the chondrogenic region of a limb from a stage-23 or stage-24 embryo is implanted into a normal growing limb bud *in ovo*, the block of cells will be incorporated into the host limb pattern, forming soft tissue in the soft tissue region of the host and cartilage in the cartilage region of the host (Searls, 1968; cf. Section I,D).

Several of the experiments mentioned above (1, 2, 3, and 4) might be explained by migration of prospective cartilage cells into the graft or by segregation of prospective cartilage cells to the region of cartilage in the graft. However, no evidence for cell movement has been obtained in the limb. When the proximal mesoderm implanted under the apical ectodermal ridge was mouse, no chick cells invaded the implant (Cairns, 1965). When the proximal mesoderm that was implanted under the apical ectodermal ridge was leg, the distal cartilaginous limb structures that were formed were leg (Saunders *et al.*, 1957, 1959). When a limb bud labeled with [^{3}H]thymidine was grafted to the stump of a limb of an unlabeled host embryo, no cell movement was observed between the unlabeled host embryo and the labeled limb (Gasseling and Saunders, 1961). When a block of limb mesenchyme cells labeled with [^{3}H]thymidine was implanted into an unlabeled limb, there was no suggestion that either the implanted cells or the host cells had moved during further development (Searls, 1967). Although dissociated limb cartilage cells will segregate from limb muscle cells or limb mesenchyme cells (Searls, 1971), this ability of cartilage cells to segregate does not appear until relatively late in cartilage differentiation. Even as late as stage 26, cells from the cartilage-forming region of the limb will not segregate from limb mesenchyme cells (Searls, 1969).

No experimental evidence has been obtained to date that suggests that limb mesenchyme cells are committed in any sense to a chondrogenic phenotype prior to stage 22, or that cartilage cells are in any sense present in the limb prior to stage 25.

B. Proximal Influences

If "influences" proximal to the limb bud itself are important in the control of cartilage differentiation, these "influences" must either act by creating a mosaic in the early stages of limb development (a possibility considered in the previous section) or act rather late in the outgrowth of the limb. The chondrogenic region of the limb does not appear until stage 22, and extracellular metachromatic material is not present in the limb until stage 25 (see Section I,A,2). The cells in the cartilage-forming region of the limb appear to be regulative until the end of stage 24; those properties of the limb that cause some limb mesenchyme cells to become cartilage and other limb mesenchyme cells to become soft tissue are fully active through stage 24 and are active to some degree until stage 26 (cf. Section I,D).

Limb buds, or the limb-forming region, from stage-15 or older embryos that have been cleaned of all nonlimb tissues will develop a normal set of limb cartilages if they are grafted to the flank (Hamburger, 1938), into the coelom (Hamburger, 1938; Eastlick, 1943; Rudnick, 1945; Pinot, 1970) or onto the chorioallantoic membrane (Murray, 1926; Hunt, 1932). The adjacent axial mesenchyme (somitic and intermediate) is required for normal limb development if the tissue grafted into the coelom is the limb-forming region from stage-11 to stage-15 embryos (Pinot, 1970). However, the limb territory cleaned of all nonlimb tissues will produce cartilage as early as stage 9, and the nature of the cartilages formed from the limb-forming regions of younger embryos (whether winglike or leglike) is controlled by the limb-forming region, not by the axial tissues (Pinot, 1970). The early requirement for the adjacent axial tissues might suggest that the axial mesenchyme is involved in the initiation of limb outgrowth. It seems clear that the axial mesenchyme adjacent to the limb bud is not involved in the control of cartilage differentiation.

It would appear to be equally clear that cartilage differentiation in the limb does not require either the invasion of the limb by nervous tissue or a pattern of vascularization. Limb buds from stage-15 or older embryos that have been cleaned of all nonlimb tissue and grafted into the coelom may attach to the mesenteries, the yolk sac, the umbilicus, or the parietal peritoneum and grow to produce a normal set of limb cartilages (Hamburger, 1938; Pinot, 1970). Limbs from stage-17 and older embryos that have been cleaned of all nonlimb tissue and grafted to the chorioallantoic membrane may also develop a normal set of limb cartilages (Murray, 1926). Thus, the invasion of the limb by nervous tissue does not seem to be required for the development of a normal set of limb cartilages (Hamburger, 1939). Limb buds from stage-18 and older embryos that have been cleaned of all nonlimb tissue and cultured on a 0.45 μm pore size filter on the chorioallantoic membrane form cartilaginous limb structures (although they were very small and highly distorted) (Searls, 1968). Similar limb buds that had been cultured on 0.22 μm pore size filters with adequate culture medium also form cartilaginous limb structures (H. G. Coon, personal communication). Since these cultured limb buds could not have been invaded by nervous tissue and clearly were not vascularized, the invasion of nervous tissue and a pattern of vascularization cannot be required for the differentiation of cartilage in a limb pattern.

C. Control by the Apical Ectodermal Ridge

Removal of the apical ectodermal ridge from the limb of an embryo stage 22 or younger will cause failure of limb outgrowth and failure of differentiation of distal cartilaginous limb structures (Saunders, 1948; Hampe, 1959; Barasa, 1960). Grafting of an apical ectodermal ridge to any other site on a limb will cause outgrowth at that site and the formation of cartilaginous limb structures (Zwilling, 1956; Saunders and Gasseling, 1968). Grafting of prospective soft tissue cells from a stage-24 limb to the region under the apical ectodermal ridge will cause the prospective soft tissue cells to form cartilaginous limb structures (Saunders *et al.*, 1957, 1959; Cairns, 1965). Fragments of a limb mesoblast or disaggregated and reaggregated limb mesoderm will form a small nodule or rod of cartilage if not placed in a limb ectodermal jacket, but will form cartilaginous limb structures if placed in contact with an apical ectodermal ridge (Zwilling, 1964). These observations suggest that the apical ectodermal ridge profoundly influences cartilage differentiation in the limb.

However, the apical ectodermal ridge does not seem to be required for the differentiation of limb long bone cartilage. If the apical ectodermal ridge is removed from a stage-19 wing bud, the wing bud will produce a humerus that is normal in size and morphology even though the more distal cartilaginous limb structures do not form. If the apical ectoderm is removed from a leg bud as early as stage 17, a femur may be obtained (3 out of 7 cases, Hampe, 1959). If the apical ectoderm is removed from a wing bud that measures 0.04 to 0.08 mm from the body wall to the apex of the wing, a humerus is usually formed and sometimes a radius and ulna (R. L. Searls, unpublished observations). The apical ectodermal ridge does not appear during outgrowth if the apical ectoderm is removed at stage 17. If the presence of a humerus has not already become somehow implicit in the limb at a very early stage (cf. Section II,A for a discussion of this possibility), the ridge cannot be held responsible for the normal humerus that develops 36 hr or more after the removal of the ridge. Milaire (1965) has suggested that "the mammalian ectodermal ridge is only concerned with the formation of the distal limb segment." The apical ectodermal ridge may be concerned only with the formation of distal cartilaginous limb structures in the chick as well.

D. Control by Physical or Nutritional Factors

1. *Anoxia*

It has been proposed that anoxia may be involved in cartilage differentiation. Bassett and Herrmann (1961) observed that cells from the tibial cortex of 20-day chick embryos formed bone when grown at high cell density and under high oxygen tension (35% O_2), cartilage when grown at high cell density and under low oxygen tension (5% O_2), and fibrous connective tissue when grown at low cell density and high oxygen tension. They suggest that the decision as to whether cells become bone, cartilage, or fibrous connective tissue might depend upon cell density and oxygen tension. Hall (1968) cultured 9-day chick cranial rudiments under conditions of normal (20% O_2) and low (5% O_2) oxygen tension and found that only bone formed under a normal atmosphere, but that 25% of the rudiments (6 out of 24) formed "precartilage" under low oxygen tension. [Rudiments cultured under a normal atmosphere that were "mechanically stimulated" formed cartilage in 66% of the rudiments (Hall, 1969).] Pawelek (1969) cultured disaggregated cartilage cells from chick sterna under normal (20% O_2) and low (5% O_2) oxygen tension and in the presence or absence of thyroxine (Table V). Whether thyroxine was present or absent, anoxia caused about a twofold increase in the rate of mucopolysaccharide synthesis. The presence or absence of thyroxine, on the other hand, had a much more profound effect on the rate of synthesis of mucopolysaccharide. Under low oxygen tension, thyroxine caused a thirty- to fortyfold stimulation in the rate of uptake of sulfate into mu-

TABLE V

Incorporation of [^{14}C]Glucose and [^{35}S]Sulfate into Polysaccharides by Chondrocyte Monolayers[a]

	+Thyroxine		−Thyroxine	
	Air	5% O_2	Air	5% O_2
Incorporation of [^{14}C]glucose	0.95	2.15	0.25	0.32
Incorporation of [^{35}S]sulfate	0.065	0.150	0.008	0.004

[a] Monolayers were placed under the described conditions after 7 days *in vitro* and tested after 9 days. Values are expressed as millimicromoles per hour per microgram of DNA. Values taken from Pawelek (1969).

copolysaccharide. It is possible that the results of Caplan (1970) may also be interpreted in terms of an induced anoxia. Caplan observed that both 3-acetylpyridine and high cell density tend to increase the number of cells from limbs stage 24 or younger that participate in the synthesis of extracellular metachromatic material. Both 3-acetylpyridine and high cell density might lead to a decrease in oxidative metabolism; the 3-acetylpyridine by being incorporated into 3-acetylpyridine DPN [which probably cannot transfer electrons to the cytochromes as efficiently as DPN itself (Kaplan et al., 1956)] and high cell density through a rapid exhaustion of the oxygen in the nutrient medium.

However, several observations suggest that the formation of the chondrogenic region during stage 22 cannot be due to anoxia in that region. During limb development the limb mesenchyme is well vascularized. By stage 19, a vascular sinus (the terminal sinus) is present a short distance under the apical ectodermal ridge. By stage 22, when the chondrogenic region is forming, other vascular sinuses are present in the limb mesenchyme more proximal than the terminal sinus. When the chondrogenic region forms, it is almost outlined by vascular sinuses (Fig. 2), and these vascular sinuses continue to be present dorsal and ventral to the chondrogenic region at least until stage 25 (cf. Barasa, 1960, for a description of the vascularization of the chick limb). When the chondrogenic region forms it does not show a gradient of cell density, of [^{35}S]sulfate uptake, or of cell division. Thus the chondrogenic region forms in proximity to an elaborate vascular system, and the differentiation of the cells in the chondrogenic region does not appear to vary with distance from the vascular system.

Further evidence that anoxia is not responsible for the differentiation of limb cartilage can be gained from culture experiments. When limb buds from stage-19 to stage-22 embryos were cultured on a 0.45 μm pore size filter on the chorioallantoic membrane, the cartilage that formed in the limb buds was in a limb pattern (indicating that cartilage did not form at random) but the cartilage was sometimes next to the filter, sometimes in the center of the cell mass, and sometimes in contact with the atmosphere. These cultures were not vascularized and appeared to be of uniform cell density until cartilage began to appear. The differentiation of cartilage in the cultures did not seem to have any constant relationship with the availability of oxygen. Medoff (1967) cultured limb mesenchyme cells from stage-19 to -20 embryos on a rotary shaker (1×10^6 cells in 5 to 9 ml of medium). These cells aggregated and produced only cartilage by the end of 6 days in culture. Since the medium was changed either every day or every other day (and presumably the cultures were gassed each time the medium was changed) and the cultures were shaken constantly, it would seem unlikely that the cultures became anaerobic. Fragments of limb bud mesoderm (about 0.1 mm thick) grown at a clot–atmosphere interface will generally form nodules of cartilage (Zwilling, 1964, 1966). It would seem unlikely that oxygen could not penetrate to the center of such a fragment.

This is not to suggest that anoxia has nothing to do with cartilage differentiation. All of the experiments described in the first paragraph of this section employed connective tissue cells except the experiment of Caplan. It is possible that anoxia stimulates the synthesis of muco-polysaccharide by cartilage cells (Hadhazy *et al.*, 1963). The extracellular metachromatic material seen at stage 25, 26, or 27 seems to be greatest in amount in the center of the cartilage-forming region, perhaps because of anoxia in that region. However, anoxia does not seem to be responsible for limb mesenchyme cells becoming cartilage cells.

2. *Nutritional Factors*

A number of chemical agents have been found to be effective in stimulating or suppressing cartilage differentiation. Among these are thyroxine (Pawelek, 1969), nicotinamide (Caplan, 1970), a high molecular weight material found in embryo extract (Coon and Cahn, 1966), and bromo-deoxyuridine (Abbott and Holtzer, 1968). One of these materials, a naturally occurring analog, or some other agent that produces a similar physiological effect, might be considered as responsible either for stimulating the synthesis of extracellular metachromatic material in the cartilage-forming region or for suppressing the synthesis of mucopolysaccharide in the prospective soft tissue region.

However, it seems unlikely that a diffusable substance provided to the limb by the vascular sinuses plays a role in cartilage differentiation (Section II,B and Section II,D,1). A gradient of differentiation or of sulfate uptake within a region of the limb is not observed (Section I,A,2 and Section I,C,1,b), thus there is no evidence that a diffusible substance with influence on cartilage differentiation is synthesized within the limb. Some agent that produces a physiological effect similar to that produced by the chemicals named above may control cartilage differentiation, but it seems unlikely that the agent is diffusible.

3. *Cell Density*

Several experiments would seem to indicate that increased cell density is important in the differentiation of cartilage. Bassett and Herrmann (1961) (Section II,D) suggested that cells cultured at high cell density become either cartilage or bone while cells cultured at low cell density become fibrous connective tissue. From the experiments of Caplan

(1970), also discussed in Section II,D, and of Umanski (1966) it is suggested that cartilage develops from limb mesenchyme cells in cell culture only when a certain cell density is exceeded. The observations of Ede and Agerbak (1968) with respect to cartilage differentiation in talpid[3] limbs also suggest that an increased cell density in the chondrogenic region is an essential stage in cartilage differentiation in the limb.

However, an increase in cell density is observed both in the chondrogenic region and in the prospective soft tissue region (Section I,A,3). The cell density in the chondrogenic region during late stage 22 is very little higher than the cell density in the prospective soft tissue region at that stage. The cell density in the prospective soft tissue region at stage 24 is higher than the cell density in the chondrogenic region during late stage 22 (Fig. 4). Thus, an increase in cell density, or an increase in cell density greater than a certain amount, does not lead directly to cartilage differentiation. In stage-19 to stage-22 limb buds cultured on a 0.45 μm pore size filter on the chorioallantoic membrane, a condensation of cells did not precede the differentiation of cartilage (Searls, 1968), and a condensation of cells did not precede the differentiation of cartilage in reaggregated limb mesenchyme cells in culture on a rotary shaker (Medoff, 1967).

E. SUMMARY

At this time, no tissue or agent has been found that exerts an effect on cartilage differentiation in the limb comparable to the effect exerted by the neural tube and the notochord on cartilage differentiation in the somite. Several possible sources of control over cartilage differentiation in the limb have been considered: (a) the limb is a mosaic system, (b) nonlimb axial tissues adjacent to the limb bud exert control over cartilage differentiaton in the limb, (c) the pattern of vascularization or of innervation exerts control over the pattern of cartilage differentiation, (d) the apical ectodermal ridge exerts control over the pattern of cartilage differentiation, (e) the physical environment of the cells in the center of the limb bud causes them to become cartilage, and (f) levels or kinds of nutrients in the center of the limb causes the cells in the center of the limb to become cartilage. None of these possiblities appears to be able to withstand analysis.

Experiments concerning limb morphogenesis (i.e., the differentiation of cartilage in a limb pattern) are generally explained either by assuming that the limb is a mosaic or by assuming that the apical ectodermal ridge establishes the pattern of the cartilages in the limb. These two theories are not distinct; if the ridge is to act at all, it must act by creating a mosaic. However, many experiments have demonstrated that the limb mesenchyme is regulative. To give only one example, if cells are taken from any region of a limb bud (including the chondrogenic region of the limb of a stage-24 embryo) and implanted in some other region of the limb bud, the cells will be integrated into the host limb pattern; the cells in the soft tissue region of the host limb will form soft tissue and the cells in the cartilage region of the host limb will form cartilage (Section I,D). This experiment also demonstrates that whatever controls whether limb mesenchyme cells become cartilage or soft tissue does not act at a particular time to create a particular territory, but must act continuously at least through stage 24. It would seem possible that the mosaic properties reside in the limb ectoderm in some manner. If the region of the limb that would normally form the radius and ulna is cut from a limb and grafted to the chorioallantoic membrane, the graft forms structures similar to those it would form *in situ* (Hampe, 1959; Kieny, 1964b); but if the ectoderm is removed from this tissue the graft will form no more than a nodule of cartilage (Zwilling, 1964, 1966). An explanation of how the limb ectoderm could have or could express mosaic properties is not obvious.

No theory capable of explaining why some limb mesenchyme cells become cartilage while other limb mesenchyme cells become soft tissue that is consonant with all of the experimental data is yet available. It is clear that much more research is necessary.

References

Abbott, J., and Holtzer, H. (1968). *Proc. Nat. Acad. Sci. U.S.* **59**, 1144–1151.
Amprino, R., and Camosso, M. (1959). *Acta Anat.* **38**, 280–288.
Amprino, R., and Camosso, M. E. (1965). *Acta Anat.* **61**, 259–288.
Amprino, R., and Camosso, M. E. (1966). *Acta Anat.* **63**, 363–387.
Barasa, A. (1960). *Riv. Biol.* **52**, 257–292.
Barasa, A. (1964). *Experientia* **20**, 443.
Bassett, C. A. L., and Herrmann, H. (1961). *Nature (London)* **194**, 460–461.
Cairns, J. M. (1965). *Develop. Biol.* **12**, 36–52.
Caplan, A. I. (1970). *Exp. Cell Res.* **62**, 341–355.
Coleman, J. R., Coleman, A. W., and Hartline, E. J. H. (1969). *Devlop. Biol.* **19**, 527–548.
Coon, H. G., and Cahn, R. D. (1966). *Science* **153**, 1116–1119.
Deuchar, E. M. (1963). *Exp. Cell Res.* **30**, 528–540.
Eastlick, H. L. (1943). *J. Exp. Zool.* **93**, 27–45.
Ede, D. A., and Agerbak, G. S. (1968). *J. Embryol. Exp. Morphol.* **20**, 81–100.
Fell, H. B., and Canti, R. G. (1934). *Proc. Roy. Soc., Ser. B* **116**, 316–349.

Franco-Browder, S., De Rydt, J., and Dorfman, A. (1963). *Proc. Nat. Acad. Sci. U.S.* **49**, 643–647.

Gasseling, M. T., and Saunders, J. W., Jr. (1961). *Develop. Biol.* **3**, 1–25.

Godman, G. C., and Porter, K. R. (1960). *J. Biophys. Biochem. Cytol.* **8**, 719–760.

Goel, S. C. (1970). *J. Embryol. Exp. Morphol.* **23**, 169–184.

Hadhazy, Cs., Olah, E. H., and Kropecher, S. (1963). *Acta Biol. (Budapest)* **14**, 67–75.

Hall, B. K. (1968). *J. Exp. Zool.* **168**, 283–306.

Hall, B. K. (1969). *Life Sci.* **8**, 553–558.

Hamburger, V. (1938). *J. Exp. Zool.* **77**, 379–397.

Hamburger, V. (1939). *J. Exp. Zool.* **80**, 347–383.

Hamburger, V., and Hamilton, H. (1951). *J. Morphol.* **88**, 49–92.

Hampe, A. (1959). *Arch. Anat. Microsc. Morphol. Exp.* **48**, 345–478.

Hay, E. D. (1968). *In* "Epithelial-Mesenchymal Interactions" (R. Billingham, ed.), pp. 31–55. Williams & Wilkins, Baltimore, Maryland.

Holtzer, H. (1968). *In* "Epithelial-Mesenchymal Interactions" (R. Billingham, ed.), pp. 152–164. Williams & Wilkins, Baltimore, Maryland.

Holtzer, H., and Detwiler, S. R. (1953). *J. Exp. Zool.* **123**, 335–365.

Holtzer, H., and Matheson, D. W. (1970). *In* "Chemistry and Molecular Biology of the Intercellular Matrix" (E. A. Balazs, ed.), Vol. 3, pp. 1753–1770. Academic Press, New York.

Hunt, E. A. (1932). *J. Exp. Zool.* **62**, 57–91.

Huxley, J. S., and de Beer, G. R. (1934). "The Elements of Experimental Embryology." Cambridge Univ. Press, London and New York.

Janners, M. Y., and Searls, R. L. (1970). *Develop. Biol.* **23**, 136–165.

Johnston, P. M., and Comar, C. L. (1957). *J. Biophys. Biochem. Cytol.* **3**, 231–238.

Jurand, A. (1965). *Proc. Roy. Soc., Ser. B* **162**, 387–405.

Kaplan, N. O., Ciotti, M. M., and Stolzenbach, F. E. (1956). *J. Biol. Chem.* **221**, 833–844.

Kieny, M. (1964a). *Arch. Anat. Micros. Morphol. Exp.* **53**, 29–44.

Kieny, M. (1964b). *Develop. Biol.* **9**, 197–229.

Kvist, T. N., and Finnegan, C. V. (1970a). *J. Exp. Zool.* **175**, 221–240.

Kvist, T. N., and Finnegan, C. V. (1970b). *J. Exp. Zool.* **175**, 241–258.

Lash, J. W. (1968a). *In* "Epithelial-Mesenchymal Interactions" (R. Billingham, ed.), pp. 165–172. Williams & Wilkins, Baltimore, Maryland.

Lash, J. W. (1968b). *In* "Stability of the Differentiated States" (H. Ursprung, ed.), pp. 17–24. Springer-Verlag, Berlin and New York.

Mathews, M. B. (1967). *Nature (London)* **213**, 1255–1256.

Medoff, J. (1967). *Develop. Biol.* **16**, 118–143.

Milaire, J. (1965). *In* "Organogenesis" (R. L. DeHaan and H. Ursprung, eds.), pp. 283–300. Holt, New York.

Mottet, N. K. (1967). *J. Exp. Zool.* **165**, 279–292.

Murray, P. D. F. (1926). *Proc. Linn. Soc. N.S.W.* **51**, 187–263.

Murray, P. D. F., and Huxley, J. S. (1925). *J. Anat.* **59**, 379–384.

O'Connell, J J., and Low, F. N. (1970). *Anat. Rec.* **167**, 425–438.

Okada, S. (1967). *J. Cell Biol.* **34**, 915–916.

Pawelek, J. M. (1969). *Develop. Biol.* **19**, 52–72.

Pinot, M. (1970). *J. Embryol. Exp. Morphol.* **23**, 109–151.

Rudnick, D. (1945). *Trans. Conn. Acad. Arts Sci.* **36**, 353–377.

Saito, H., Kamagata, T., and Suzuki, S. (1968). *J. Biol. Chem.* **243**, 1536–1542.

Saunders, J. W., Jr. (1948). *J. Exp. Zool.* **108**, 363–403.

Saunders, J. W., Jr., and Gasseling, M. T. (1968). *In* "Epithelial-Mesenchymal Interactions" (R. Billingham, ed.), pp. 78–97. Williams & Wilkins, Baltimore, Maryland.

Saunders, J. W., Jr., Cairns, J. M., and Gasseling, M. T. (1957). *J. Morphol.* **101**, 57–88.

Saunders, J. W., Jr., Gasseling, M. T., and Cairns, J. M. (1959). *Develop. Biol.* **1**, 281–301.

Saunders, J. W., Jr., Heinkel, D., Gawlik, S., and Gasseling, M. T. (1964). *Amer. Zool.* **4**, 303.

Searls, R. L. (1965a). *Develop. Biol.* **11**, 155–168.

Searls, R. L. (1965b). *Proc. Soc. Exp. Biol. Med.* **118**, 1172–1176.

Searls, R. L. (1967). *J. Exp. Zool.* **166**, 39–50.

Searls, R. L. (1968). *Develop. Biol.* **17**, 382–399.

Searls, R. L. (1969). *J. Cell Biol.* **43**, 126a.

Searls, R. L. (1971). *Exp. Cell Res.* **64**, 163–169.

Searls, R. L., and Janners, M. Y. (1969). *J. Exp. Zool.* **170**, 365–376.

Selby, D., and Murray, P. D. F. (1928). *Aust. J. Exp. Biol. Med. Sci.* **5**, 181–188.

Strudel, G. (1953). L'influence morphogene du tube nerveux sur la colonne vertebrale. *Comp. Rend. Soc. Biol.* **147**, 132–133.

Umanski, R. (1966). *Develop. Biol.* **13**, 31–56.

Watterson, R. I., I. Fowler, and B. J. Fowler (1954). The role of neural tube and notochord in development of the axial skeleton of the chick. *Amer. J. Anat.* **95**, 337–400.

Zwilling, E. (1956). *J. Exp. Zool.* **132**, 173–186.

Zwilling, E. (1964). *Develop. Biol.* **9**, 20–37.

Zwilling, E. (1966). *Ann. Med. Exp. Biol. Fenn.* **44**, 134–139.

Author Index

Numbers in italics refer to the pages on which the complete references are listed.

A

Abbott, J., 247, *249*
Abeles, F. B., 65, 66, 67, *80*
Abelson, R., 118, *153*
Abrams, M., 74, *82*
Abramsky, T., 133, *146*
Addicott, F. T., 41, *47*
Adler, J., 122, *148*
Agerbak, G. S., 248, *249*
Ainsworth, G. C., 128, *146*
Aitkhozhin, M. A., 4, *20*
Albanese, I., 10, *22*
Aldrich, H. C., 99, 104, *105*
Alff-Steinberger, C., 108, *154*
Allen, E. R., 177, 182, 186, *190*, 199, *216*
Allfrey, V. G., 2, *20*
Ames, B. N., 37, *48*
Amprino, R., 240, 241, *249*
Anderson, F., 31, *47*
Anderson, J. M., 52, 55, 68, *80, 82*
Anderson, M. B., 71, *80*
Anderson, W., *81*
Antonie, D., 58, *81*
Arias, L., 142, *154*
Armstrong, D. J., 31, *48*, 68, 70, 71, 72, *80, 83*
Aronson, A. I., 8, *20*
Aronson, J. M., 123, *146*
Ashworth, J. M., 92, *106*
Asmus, A., 122, *153*
Atwood, K. C., 115, 117, *153*
Azhar, S., 79, *80*

B

Bacon, J. S. D., 127, *146*
Badman, W. S., 87, 92, 94, 96, 102, 103, 104, *105*
Balassa, G., 123, *146*

Balz, H. P., 68, *80*
Barasa, A., 241, 244, 246, *249*
Barbata, G., 14, *20*
Barber, J. T., 128, *146*
Barbesgaard, P., 128, *146*
Barkley, D. S., 88, 89, 98, 100, 104, *105, 106*
Barkley, G. M., 58, *80*
Barnett, N. M., 57, 58, *82*
Barratt, R. W., 124, 132, *146, 154,* 189, *191*
Bartnicki-Garcia, S., 123, 124, 125, 127, *146, 151, 154*
Bashey, R. I., 208, *217*
Basl, M., 128, 131, *153*
Bassett, C. A. L., 245, 247, *249*
Bautz, A. F., 77, *80*
Bautz, E. K. F., 77, *80*
Bautz, F. A., 77, *80*
Beadle, G. W., 119, *146*
Beckett, E. B., 189, *190*
Beckwith, J., 78, *83*
Beevers, L., 55, 67, *80*
Belitsina, N. V., 4, *20*
Bell, E., 16, 17, *21*
Bennett, D. C., 129, *152*
Bennett, M. V. L., 176, *190*
Bennett, T., 213, *216*
Berg, W., 17, *20*
Bergman, K., 128, *146*
Bergman, R. A., 173, 182, *190*
Bernstein, R. L., 142, *147*
Bertsch, L. L., 123, *152*
Betz, E. H., 173, *190*
Beug, H., 89, *105*
Bickel, D., 181, *190*
Bienengräber, V., 118, *147*
Birnstiel, M. L., 6, 12, *21*
Bischoff, R., 174, 181, *190*, 197, *216*
Biswas, B. B., 77, *82*

Subject Index

A

Abscisic acid, 41, 42, 44, 46, 50, 60, 63, 78
Abscission, 65
Actin, 176, 177, 182
Actinomycin D, 13, 38, 39, 42, 56, 58, 63, 65, 66
Acrasin, 88, 100
Acridine orange, 92
Aleurone, 55, 61, 63, 64
Algae, 155–166
AMP, cyclic, 78, 79, 88, 100, 101
Amylase, 61, 62
Amyloplasts, 44
Anoxia, 245
Anthesis, 43
Anticodon, 69, 70
Antigens, surface, 89
Apical ectodermal ridge, 241, 244
Apical growth, 126, 135, 142
Ascaris, 5
Aspergillus, 125, 126
Assembly, 142
Assembly, extracellular, 124
Auxin, 49, 51, 53, 54, 73, 76, 79
Auxotrophs, 131

B

Basement lamina, 197
Benzyladenine, 71, 72
Branching, 126, 129, 135, 139, 142
Budding, 127

C

Callus, 51, 71, 76
Cartilage, 219–249
Catecholamines, 209
Cell adhesion, 100
Cell association, 99, 100
Cell contacts, 89
Cell density, 247
Cell elongation, 53, 55, 56, 59
Cell fusion, 156
Cell membrane, 59, 99, 171
Cell movement, 232
Cell junctions, 176, 188, 194, 196, 197

Cell sorting, 91
Cell walls, 107–145
  bacterial, 121
  fungal, 123
Cellulase, 67
Chitin, 124, 125, 127, 134, 142
Chlamydomonas, 156
Chloroplast, 29, 32, 35
Choline, 131, 132
Cholinesterase, 188
Chondroblasts, 196
Chondroitin sulfate, 207, 209, 230, 231, 234
Cleavage, 7, 10, 17, 161, 165
Coleoptiles, 58
Collagen, 173, 208, 211, 215, 236
Colony formation, 156, 159, 160
Compartmentalization, 138
Conidia, 130
Cotyledons, 24, 26, 32, 37, 38, 46
Cycloheximide, 38, 56, 57, 65, 66
Cytokinins, 31, 49, 54, 60, 68, 70, 71, 72, 79

D

Dedifferentiation, 99
2,4-Dichlorophenoxyacetic acid, 51
Dictyostelium, 85–106, 113
Deoxyribonucleic acid
  chloroplast, 30
  mitochondrial, 5
  synthesis, 205
DNase, 68
DNA polymerase, 199, 211
Drosophila, 113, 115, 117

E

Endocardium, 197
Endoplasmic reticulum, 65
Endosperm, 24
Endosperm absorption, 25
Environmental factors, 116, 160, 164
Enzyme deficiencies, primary, 131
Epicotyl, 58
Ethylene, 50, 65
Eudorina, 160
Extracellular matrix, 206, 207

264